AF766553

Bridge Engineering - Recent Advances and Applications

Edited by Salih Yilmaz and Yavuz Yardim

Published in London, United Kingdom

Bridge Engineering - Recent Advances and Applications
http://dx.doi.org/10.5772/intechopen.107732
Edited by Salih Yilmaz and Yavuz Yardim

Contributors
Antonio Jaramillo-Morilla, Antonio Jesús Torralba-Silgado, Álvaro Serrano-Chacón, Christian Paglia, Eduardo Hidalgo-Fort, Emilio J. Mascort-Albea, Fernando Muñoz Chavero, Hayder A. Rasheed, Husam A. Alshareef, Khalid W. Al Shboul, Pedro Blanco-Carmona, Ramon Gonzalez-Carvajal, Salih Yilmaz, Vladimir Križaić, Yavuz Yardim

© The Editor(s) and the Author(s) 2024
The rights of the editor(s) and the author(s) have been asserted in accordance with the Copyright, Designs and Patents Act 1988. All rights to the book as a whole are reserved by INTECHOPEN LIMITED. The book as a whole (compilation) cannot be reproduced, distributed or used for commercial or non-commercial purposes without INTECHOPEN LIMITED's written permission. Enquiries concerning the use of the book should be directed to INTECHOPEN LIMITED rights and permissions department (permissions@intechopen.com).
Violations are liable to prosecution under the governing Copyright Law.

(cc) BY

Individual chapters of this publication are distributed under the terms of the Creative Commons Attribution 3.0 Unported License which permits commercial use, distribution and reproduction of the individual chapters, provided the original author(s) and source publication are appropriately acknowledged. If so indicated, certain images may not be included under the Creative Commons license. In such cases users will need to obtain permission from the license holder to reproduce the material. More details and guidelines concerning content reuse and adaptation can be found at http://www.intechopen.com/copyright-policy.html.

Notice
Statements and opinions expressed in the chapters are these of the individual contributors and not necessarily those of the editors or publisher. No responsibility is accepted for the accuracy of information contained in the published chapters. The publisher assumes no responsibility for any damage or injury to persons or property arising out of the use of any materials, instructions, methods or ideas contained in the book.

First published in London, United Kingdom, 2024 by IntechOpen
IntechOpen is the global imprint of INTECHOPEN LIMITED, registered in England and Wales, registration number: 11086078, 167-169 Great Portland Street, London, W1W 5PF, United Kingdom

British Library Cataloguing-in-Publication Data
A catalogue record for this book is available from the British Library

Additional hard and PDF copies can be obtained from orders@intechopen.com

Bridge Engineering - Recent Advances and Applications
Edited by Salih Yilmaz and Yavuz Yardim
p. cm.

This title is part of the Civil Engineering Book Series, Volume 9
Topic: Structural Engineering
Series Editor: Assed Haddad
Topic Editor: Wael A. Zatar

Print ISBN 978-1-83769-005-3
Online ISBN 978-1-83769-006-0
eBook (PDF) ISBN 978-1-83769-007-7
ISSN 3029-0287

We are IntechOpen, the world's leading publisher of Open Access books

Built by scientists, for scientists

7,100+
Open access books available

189,000+
International authors and editors

205M+
Downloads

156
Countries delivered to

Our authors are among the
Top 1%
most cited scientists

12.2%
Contributors from top 500 universities

CLARIVATE ANALYTICS
BOOK CITATION INDEX
INDEXED

WEB OF SCIENCE™

Selection of our books indexed in the Book Citation Index in Web of Science™ Core Collection (BKCI)

Interested in publishing with us?
Contact book.department@intechopen.com

Numbers displayed above are based on latest data collected.
For more information visit www.intechopen.com

For EU product safety concerns:
IN TECH d.o.o., Prolaz Marije Krucifikse Kozulić 3, 51000 Rijeka, Croatia,
info@intechopen.com or visit our website at intechopen.com.

IntechOpen Book Series
Civil Engineering
Volume 9

Aims and Scope of the Series

Civil engineering is a traditional field of engineering from which most other branches of engineering have evolved. It comprises traditional sub-areas like transportation, structures, construction, geotechnics, water resources, and building materials. It also encompasses sustainability, risk, environment, and other concepts at its core. Historically, developments in civil engineering included traditional aspects of architecture and urban planning as well as practical applications from the construction industry. Most recently, many elements evolved from other fields of knowledge and topics like simulation, optimization, and decision science have been researched and applied to increase and evolve concepts and applications in this field. Civil engineering has evolved in the last years due to the demands of society in terms of the quality of its products, modern applications, official requirements, and cost and schedule restrictions. This series addresses real-life problems and applications of civil engineering and presents recent, cutting-edge research as well as traditional knowledge along with real-world examples of developments in the field.

Meet the Series Editor

Professor Assed N. Haddad is a Civil Engineer with a degree from the Federal University of Rio de Janeiro (UFRJ) earned in 1986, as well as a Juris Doctor degree from the Fluminense University Center earned in 1993, and a Master's degree in Civil Engineering from the Fluminense Federal University (UFF) obtained in 1992. He completed his Ph.D. in Production Engineering from COPPE / Federal University of Rio de Janeiro in 1996. Professor Haddad's academic pursuits have taken him to postdoctoral stays at the University of Florida, USA in 2006; at the Universitat Politècnica de Catalunya, Spain in 2010; and at the University of New South Wales Sydney, Australia in 2019. Currently, he serves as a Full Professor at the Federal University of Rio de Janeiro. He has held visiting professorships at various institutions including the University of Florida, Universitat Politècnica de Catalunya, Universitat Rovira i Virgili, and Western Sydney University. His research expertise encompasses Civil, Environmental, and Production Engineering, with a primary focus on the following topics: Construction Engineering and Management, Risk Management, and Life Cycle Assessment. He has been the recipient of research grants from the State of Rio de Janeiro, Brazil: CNE FAPERJ from 2019 to 2022 and from 2023 to 2025. Additionally, his research grants obtained from the Brazilian Government CNP since 2012 last to this date. Professor Haddad has been involved in several academic endeavors, being the Guest Editor of the International Journal of Construction Management; MDPI's Sustainability, Energies, and Infrastructures; Associate Editor at Frontiers in Built Environment / Sustainable Design and Construction; Guest Editor at Frontiers in Built Environment / Construction Management; and Academic Editor of the Journal of Engineering, Civil Engineering Section of Hindawi. He is currently a Professor of the Environmental Engineering Program at UFRJ and the Civil Engineering Program at UFF.

Meet the Volume Editors

Dr. Salih Yilmaz received his Ph.D. in civil engineering from Pamukkale University, Turkey, in 2007. After serving as a research assistant and lecturer there, Dr. Yilmaz transferred to Katip Celebi University, Turkey, where he founded the Department of Civil Engineering and Earthquake Engineering Research Center in 2012 and 2014, respectively. He was a faculty member at both Katip Celebi University and Pamukkale University and a research associate at the University of Texas at Austin, USA. Dr. Yılmaz is involved in several national and international research projects and has consulted for many companies. He worked as a senior structural engineer. His interests include structural damages, seismic performance of existing buildings, and seismic retrofit and strengthening of structures. He completed strengthening projects and consultancy over 1.000.000 m2 of total building area. Currently, he is working at Eastern Switzerland University of Applied Sciences.

Dr. Yavuz Yardım, holding a Ph.D. in Structural Engineering and recognized as a CEng MICE in the UK, has served in academic roles across Malaysia, Albania, Oman, Turkey, and the UK. Currently, Dr. Yardım is part of the Civil and Environmental Engineering discipline at the University of Edinburgh. His significant contributions include developing an expert system application for bridge inspectors in Malaysia and leading the inspections of over 150 significant bridges in Albania. He has authored over 70 peer-reviewed articles in structural engineering. Dr. Yardım's research experience and areas of interest encompass the experimental and numerical investigation of reinforced concrete structures, earthquake-resistant buildings, and the performance of historical masonry structures and RC bridges. He also has extensive experience in predicting the behavior of unreinforced masonry and RC buildings and historical structures under static and dynamic loads. Dr. Yardım has also been involved in and led several academic projects.

Contents

Preface **XIII**

Section 1
Bridge Engineering Fundamentals **1**

Chapter 1 **3**
Introductory Chapter: Challenges and Advancements in Bridge Engineering
by Salih Yilmaz and Yavuz Yardim

Chapter 2 **7**
The Durability of Reinforced Concrete Bridges
by Christian Paglia

Chapter 3 **27**
Advanced Analytical Methods for Fatigue Assessment of Ancillary Systems in Highway Bridges
by Khalid W. Al Shboul, Husam A. Alshareef and Hayder A. Rasheed

Section 2
Bridge Monitoring and Maintenance **51**

Chapter 4 **53**
Planning and Repairing Bridges with Robotic and Automation Technology
by Vladimir Križaić

Chapter 5 **75**
IoT Framework Application to SHM Systems
by Eduardo Hidalgo-Fort, Pedro Blanco-Carmona, Alvaro Serrano-Chacon, Emilio José Mascort-Albea, Fernando Muñoz-Chavero, Ramón González-Carvajal, Antonio Jesús Torralba-Silgado and Antonio Jaramillo-Morilla

Preface

Bridges are more than just physical structures; they are symbols of human ingenuity and progress. Throughout history, they have connected communities, facilitated trade, and shaped the landscape of civilizations. The need to cross obstacles likely spurred the creation of bridges, perhaps even before humans built temporary shelters. From the humble beginnings of logs laid across streams to the awe-inspiring suspension bridges of today, bridge engineering has undergone a remarkable evolution.

The modern era of bridge engineering is characterized by a surge in innovation. New materials, advanced analytical methods, and cutting-edge technologies have revolutionized the field. These advancements have led to the construction of bridges that are not only aesthetically pleasing but also safer, more durable, and more sustainable.

However, significant challenges lie ahead. Aging infrastructure, increasing traffic loads, and the effects of climate change pose serious threats to the longevity and safety of our bridges.

This edited volume aims to address these challenges by providing a comprehensive overview of recent advances and applications in bridge engineering. The book explores strategies for extending the lifespan of existing bridges, examines innovative materials and construction techniques that promote sustainability, and delves into cutting-edge methods for assessing bridge fatigue. Additionally, it takes a forward-looking approach, highlighting the importance of ongoing bridge preservation to ensure safety and functionality, exploring the latest advancements in bridge planning, repair, and monitoring through robotics, automation, and the potential of artificial intelligence.

This book is a valuable resource for a broad audience, including researchers exploring new frontiers, practitioners seeking the latest advancements, policymakers developing effective strategies, and students gaining a comprehensive understanding of the field.

By showcasing recent advancements and applications, this book equips bridge engineers with the knowledge and tools to design more efficient, resilient, and sustainable structures for the future. We believe this edited volume will be a valuable resource for all stakeholders in the field, fostering innovation and advancing the art of bridge design.

A sincere thanks to all the contributing authors who have shared their expertise and insights, making this book a comprehensive and valuable resource.

Sincerely,

Salih Yilmaz
Institute for Civil and Environmental Engineering,
University of Applied Sciences,
Rapperswil, Switzerland

Yavuz Yardim
Department of Civil and Environmental Engineering,
The University of Edinburgh,
Edinburgh, UK

Section 1

Bridge Engineering Fundamentals

Chapter 1

Introductory Chapter: Challenges and Advancements in Bridge Engineering

Salih Yilmaz and Yavuz Yardim

1. Introduction

Throughout history, bridges have served as vital arteries in the human network, connecting communities, fostering trade, and facilitating cultural exchange. They are more than just functional structures; they are testaments to human ingenuity and progress, reflecting the social, economic, and cultural realities of their time [1]. From the humble beginnings of logs laid across streams, bridge design has undergone a remarkable evolution. Each era has left its mark, contributing a unique blend of form and function.

The earliest bridges were likely simple log structures, used to traverse small streams and waterways. As civilizations developed and transportation needs grew more complex, bridge design became more sophisticated. The ancient Egyptians, for example, constructed stone beam bridges, some of which remain standing today [2]. The Romans further advanced bridge construction with their innovative use of arches. Roman arch bridges, such as the Pont du Gard in France, showcased remarkable precision and craftsmanship and continue to inspire us today with their enduring resilience [3].

2. A journey through bridge engineering

The Middle Ages saw the development of truss bridges, which utilized a network of wooden beams to distribute weight across the span. These bridges were particularly well-suited for longer distances and were widely used throughout Europe [2]. The Industrial Revolution ushered in a new era of bridge construction, with the introduction of new materials like iron and steel. Suspension bridges, such as the iconic Menai Suspension Bridge in Wales, became possible thanks to these advancements [4].

The 20th century witnessed further innovations in bridge engineering, with the development of prestressed concrete and cable-stayed bridges. Prestressed concrete bridges offer improved durability and strength compared to traditional concrete, while cable-stayed bridges offer a more esthetically pleasing design with their single tower and radiating cables [5]. Today, bridge engineers have a vast array of materials and techniques at their disposal, allowing them to create ever-more sophisticated and impressive structures.

IntechOpen

3. The current landscape of bridge engineering

The field of bridge engineering is experiencing a period of profound transformation. New materials, such as high-performance concrete and fiber-reinforced polymers, are offering improved strength, durability, and sustainability [6]. Advanced technologies, like digital modeling and 3D printing, are revolutionizing the way bridges are designed and constructed [7]. Innovative analytical methods, such as finite element analysis, are allowing engineers to more accurately predict the behavior of bridges under different loads [5].

These advancements enable us to create bridges that are not only esthetically pleasing but also safer, more durable, and more sustainable. Bridges are no longer solely utilitarian structures; they can be designed to integrate seamlessly into the surrounding environment and even generate renewable energy through solar panels or wind turbines [8]. They stand as symbols of progress and human achievement, showcasing our capacity to overcome engineering challenges and connect people across vast distances.

4. Challenges and opportunities for the future

Despite these remarkable strides, bridge engineering faces significant challenges. Aging infrastructure is a pressing concern, with many bridges around the world nearing the end of their lifespan [9]. Maintaining and rehabilitating these structures presents complex technical and financial hurdles. Traditional inspection methods can be time-consuming and labor-intensive, and there is a growing need for more efficient and cost-effective ways to assess bridge health [7].

Additionally, increasing traffic volumes and heavier vehicles place greater demands on these aging structures. As cities grow and populations expand, our infrastructure needs to adapt. Designing and building bridges capable of handling these increased loads is a crucial challenge for the future [5].

Climate change and extreme weather events pose further threats. Floods, hurricanes, and other natural disasters accelerate wear and tear on bridges, adding to the complexities faced by bridge engineers [10]. Looking ahead, bridge design will need to take into account these environmental factors, ensuring new structures can withstand the rigors of a changing climate. Integrating climate resilience into bridge design will be crucial for protecting infrastructure and ensuring the safety of communities [5].

5. Research and development needs in bridge engineering

The future of bridge engineering hinges on continuous research and development (R&D) efforts focused on addressing critical challenges and capitalizing on emerging opportunities. Sustainability, safety, and durability are the cornerstones of successful bridge design and construction. This subchapter explores these crucial areas and identifies key R&D imperatives:

Sustainability: To minimize environmental impact, R&D efforts should focus on developing and implementing.

Advanced materials: Sustainable materials with lower embodied energy and carbon footprints, such as high-performance concrete with recycled aggregates, bio-based composites, and self-healing materials.

Life-cycle assessment (LCA): Standardized LCA tools and methodologies to evaluate a bridge's environmental impact throughout its lifespan.

Energy efficiency: Integration of renewable energy generation technologies, such as solar panels or wind turbines, into bridge structures.

Safety: To enhance bridge safety, R&D should prioritize.

Advanced monitoring systems: Cost-effective and reliable sensor systems for continuous structural health assessment, allowing for preventative maintenance and early detection of potential failures.

Bridge Load Modeling: Accurate load modeling that incorporates real-time traffic data and accounts for the increasing weight and volume of vehicles.

Resilient design for extreme events: Design guidelines and materials specifically tailored for enhanced bridge resilience against floods, hurricanes, and earthquakes.

Durability: To extend bridge lifespan, R&D should explore.

Innovative repair and rehabilitation techniques: Cost-effective methods for bridge repair, utilizing advanced materials and technologies like self-healing concrete or robotics.

Non-destructive testing (NDT) methods: Refined NDT methods for faster and more cost-effective bridge inspections that avoid structural damage.

Durable Materials with Improved Performance: New materials with enhanced durability and resistance to environmental degradation, such as corrosion or fatigue.

By addressing these R&D imperatives and prioritizing sustainability, safety, and durability, bridge engineering can ensure the development of reliable, resilient, and environmentally conscious bridges for future generations.

Author details

Salih Yilmaz* and Yavuz Yardim

1 Institute for Civil and Environmental Engineering, University of Applied Sciences, Rapperswil, Switzerland

2 Department of Civil and Environmental Engineering, The University of Edinburgh, Edinburgh, UK

*Address all correspondence to: thesalihyilmaz@gmail.com

IntechOpen

© 2024 The Author(s). Licensee IntechOpen. This chapter is distributed under the terms of the Creative Commons Attribution License (http://creativecommons.org/licenses/by/3.0), which permits unrestricted use, distribution, and reproduction in any medium, provided the original work is properly cited.

References

[1] Gonzalez A, Schorr M, Valdez B, Mungaray A. Bridges: Structures and Materials, Ancient and Modern. IntechOpen; 2020. DOI: 10.5772/intechopen.90718

[2] Rostem OR. Bridges in ancient Egypt with a report on a newly excavated bridge from the Old Kingdom, Giza. Annales du Service des Antiquités de l'Égypte. 1948;**48**:159-162

[3] Babic M. Ancient Roman bridges and their social significance. Acta Antiqua Academiae Scientiarum Hungaricae. 2009;**53**(1):121-132

[4] Gilbert D. On the mathematical theory of suspension bridges, with tables for facilitating their construction. Philosophical Transactions of the Royal Society of London. 1826;**116**(1/3):202-218

[5] American Trade Initiatives, Inc., & Avalon Integrated Services, Inc. Performance of concrete segmental and cable-stayed bridges in Europe. Washington DC, USA: U.S. Department of Transportation, Federal Highway Administration; 2001

[6] Lantsoght EOL, editor. Advanced structural concrete materials in bridges. Heidelberg, Germany: Springer; 2023

[7] Gardner L, Kyvelou P, Herbert G, Buchanan C. Testing and initial verification of the world's first metal 3D printed bridge. Automation in Construction. 2020;**172**:106233

[8] Soto Hernández Ó, Volkov K, Martín Mederos AC, Medina Padrón JF, Feijóo Lorenzo AE. Power output of a wind turbine installed in an already existing viaduct. Renewable and Sustainable Energy Reviews. 2015;**48**:287-299

[9] U.S. Department of Transportation, Federal Highway Administration & Federal Transit Administration. Status of the Nation's Highways, Bridges, and Transit: Conditions & Performance Report to Congress. 24th ed. Washington, DC: U.S. Department of Transportation; 2021

[10] Schultz AE, Gastineau AJ. Bridge collapse. In: Pipinato A, editor. Innovative bridge design handbook. New York, NY, USA: Butterworth-Heinemann; 2016. pp. 795-815

Chapter 2

The Durability of Reinforced Concrete Bridges

Christian Paglia

Abstract

Bridge materials durability is a key issue to be considered for reinforced concrete bridges. The maintenance and the repair are critical and costly. These issues may exhibit significant implications for the safety and functionality of these transportation infrastructures. Reinforced concrete bridges must withstand atmospheric environmental and artificial service conditions. They have to resist the adverse effects of cracking, carbonation, chloride migration, steel rebar corrosion, as well as wind, snow, rain, and freezing actions. The cyclic variation of these conditions has a relevant consequence on the long-term resistance. The durability of reinforced concrete bridges is also influenced by the geographic orientation and the local environmental exposition of the several bridge elements. The upper and lower structures of the bridges are linked through special bearings, which exhibit degradation over time. Foundation piles must be monitored with respect to the conservation state. Therefore, all durability aspects must be considered prior to planning and repairing the bridges. In this manner, it is possible to design bridges with specific needs and ensure the service endurance and safety.

Keywords: bridges, durability, concrete, rebars, environmental degradation

1. Introduction

The reinforced concrete bridges are widely widespread in the world. They are mainly composed of three fundaments materials: rock aggregates, cementitious binders, and steel rebars. The rock aggregates and the cement-based components form the main ingredients for the concrete. The concrete is mainly responsible for the compressive strength solicitation of the infrastructure. The addition of steel rebars mainly enables to whole structure to withstand the tensile stresses. From a material point of view, the combination of steel rebars and concrete forms a composite material capable of facing some of the main solicitation stresses that arise during the service life of bridges. This composite material, namely the reinforced concrete, must give a fundamental contribution to the quality and durability of the infrastructure. Each type of component needs to be chosen with attention. The rock aggregates must exhibit a high quality, adequate lithology, hardness, mineralogy, resistance to fragmentation, shape, and form, and the combination of the aggregate size must follow well-defined granulometric curve ranges [1] in order to be used for the production of concrete. On the other hand, the choice of the cement type also needs to accomplish clear parameters to satisfy the strength development requirement, the bonding capability, and the durability of the concrete [2, 3]. Today, also the sustainability plays a major role. The steel rebars are also required to fit within specified requirements concerning the mechanical properties, fatigue, weldability, and the geometrical factors [4]. The adhesion bond between the steel rebars and the concrete is a major concern with respect to the durability of the entire construction system. The deterioration of this latter building system is affected by the mechanical solicitations and from the environmental exposure condition. The atmospheric interaction causes carbonation, chloride migration, freezing action, cracking and scaling of the concrete, and corrosion of the steel rebars [3].

The conventional use of type of cement, in particular, CEM I-based Ordinary Portland cement, is a common practice. Various strength class cements are used, ranging from 32.5 to 52.5 Mpa with a normal (N) or rapid (R) development of the strength. Nevertheless, the CO_2 development associated with the production of types of cement and the general sustainability target to reduce such emission, triggered the use of alternative binders or recycled components. The addition of supplementary cementitious materials with filler or hydraulic properties or a combination of them is largely investigated [5], while the substitution or addition of recycled components to the Ordinary Portland cement component is sometimes largely disregarded. The long-term service life requirements of bridges up to 80–100 years, still arise some questions related to the low material quality of recycled concrete for infrastructures.

It must be stated that nowadays, the compressive strength target up to 35 MPa can be easily reached. Nevertheless, the durability still remains an issue to be solved. The high durability requirements asked through laboratory tests [3], concerning the freeze/thaw resistance, the depth of carbonation, and the chloride diffusion tend to decrease the permeability of the material at relatively early age up to 28 days. Consequently, high modulus of elasticity and stiffness are seen. The structure may become more rigid and more prone to micro cracking. The latter feature significantly lowers the service life. That means high strength does not mean necessarily longer service life. Furthermore, a higher durability may be achieved with more sustainable materials and combination of them to produce safe and more durable reinforced concrete bridges.

2. Concrete degradation

2.1 General deterioration

The concrete is a very versatile building material. The technology development of the last decades allowed to obtain complex structures and to speed up the casting procedure. On the other hand, cementitious materials need a special attention. High compressive strength and modulus of elasticity, as well as the load intensity and fatigue, may cause crack formation (**Figure 1** left). The cracks can be seen with the bare eye, by spraying water on the surface and detecting the darker areas along the cracks. These latter may exhibit a fine widespread pattern or a sequence with variable opening up to a few millimeters. The multiple material composition causes a relatively wide susceptibility to several environmental actions, while chemical interactions between the hydrated cement pastes with the aggregates or the physical/chemical reaction between the cementitious matrix and the external elements such as CO_2, chloride, acids, water leaching of the cement paste due to the mechanical droplet

Figure 1.
Bridges across a highly trafficked highway in service for over 40 years. Cracking (left), coloring effects (center-left), dirt accumulation (center-right), and segregation (right).

actions transported by the vehicles, the windy conditions, and the temperature variation all contribute to damage. Alkali-silica reaction between the amorphous Silicon-rich aggregates and the cement alkalis form an expansive gel and lead to multiple chaotic net-like cracking [6]. The contact between sulfate-rich waters and the aluminate content of the cement leads to the formation of expansive hexagonal-shaped ettringite crystals, which lead to expansion and cracking. This is particularly observed for Na_2SO_4, while with $MgSO_4$ a disgregation is seen [7]. Acids strongly attack the basicity of the concrete and cause a relevant dissolution of the material and in some cases directly the steel rebars [8]. Nonetheless, the degrading phenomena occurring along bridges are load-related microcrack formation, temperature-induced thermal expansion, water leaching, and freezing, [3, 9] and the main focus must be driven to carbonation and chloride contamination.

2.2 The concrete carbonation

The cement hydration produces portlandite $Ca(OH)_2$, which is responsible for the high alkalinity of the concrete pore solution. The basicity ranges at pH between 13 and 13.5 [10]. The concrete of bridges is constantly in contact with the CO_2 of the atmosphere. The CO_2 diffusion within the concrete causes the reaction with the portlandite to form calcium carbonate. The reaction depends on the humidity [11], the $Ca(OH)_2$ availability, the porosity, and the microstructure [12, 13]. The carbonation takes place at a relatively constant rate [14]. The process does not directly damage the concrete. On the contrary, the precipitation of carbonate may be seen as a partial carbon sink for the environment, and the $CaCO_3$ formation may, in some cases, close the porosity at an initial stage. This may lower the further CO_2 penetration within the concrete. Nevertheless, in the field along bridges, this does not take place as expected. Additional detrimental atmospheric effects allow the CO_2 to continuously enter the microstructure. The carbonation of the concrete may have a slight coloring effect on the concrete surface. It becomes whitish as compared to the intense gray color of the as-cast in place concrete (**Figure 1** left-center). In some areas, especially if the bridges cross highly trafficked highways, a significant deposition of dust and dirt is seen along many parts of the structures, and the surfaces become covered with a dark-colored deposition (**Figure 1** center-right). It appears that such a natural layer of deposits does not significantly influence the CO_2 penetration. In spite of the high-quality concrete that is transported on-site, the cast and placing, play a relevant role in the long-term duration of the infrastructure. This is not always an easy issue and locally, segregations (**Figure 1** right), delamination, or surface detachments caused by an inadequate formwork placing, and treatment may accelerate the entrance of the CO_2. This may be accounted for one of the main partial lack of direct correlation between laboratory accelerated carbonation tests [3] and real field behavior.

3. Steel rebar corrosion

The corrosion of steel rebars takes place within concrete when the alkalinity is lowered from pH 13.5 to about pH 9. Under these conditions, the passive protective layer on the rebars is no longer stable and corrosion initiates [14–16]. The iron hydroxide corrosion products exhibit a higher volume as compared to the original iron in steel. An expansion action occurs. This leads to high local stresses along the rebars and cracks form. The formation of cracks further accelerates the penetration of water, and the corrosion is promoted. The carbonation solely does not adversely affect the steel rebars or the concrete durability. On the other hand, the combination of carbonation, the lowering of the pH, and the presence of water leads to the steel rebar corrosion. This type of corrosion is called uniform corrosion and is uniformly widespread along the rebar surface, thus causing spalling of the concrete cover. This is observed at advanced corrosion stage. It is important to notice that the cracking does not occur only along the steel rebars but also on the sides up to 10–15 cm away from the rebar (**Figure 2** left). The tensions that arise during the corrosion process are quite widespread around the rebar regions.

The corrosion induced by chlorides that comes from the NaCl salt spreading during the winter time to avoid the water freezing along the roadways is locally more intense. It is generally enhanced by alternating wetting and drying cycles [17, 18]. The localized corrosion starts with the formation of local pits (**Figure 2** right). This form of localized corrosion of the steel rebar can lead to a reduction of the rebar diameter. This fact may cause structural problems. This form of corrosion is mainly present along the alpine spaces and in weather conditions, where an alternating seasonal temperature range between 10 and 30°C is present. In very hot climas and in the absence of freezing or in very cold regions, where the temperature may well be below −15°C, this issue is less present.

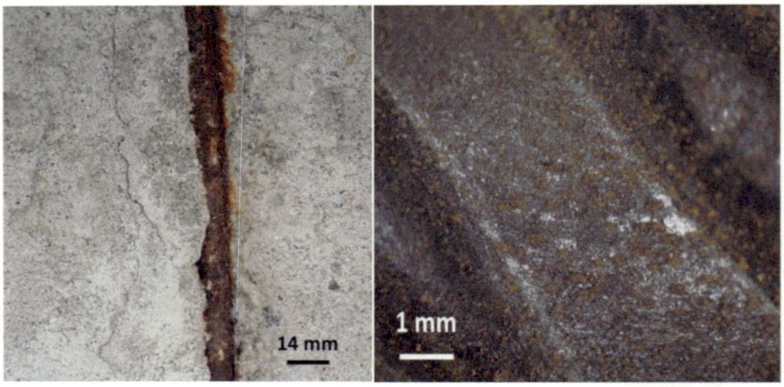

Figure 2.
Uniform corrosion induced by carbonation (left) of a bridge in service for over 40 years in a south alpine region and localized pits initiation during chloride attack (right) on a steel rebar.

4. The environmental exposure

The several parts of bridges are differently exposed to environmental conditions. Therefore, it is necessary to clearly identify the sampling areas in order to get a complete a reliable conservation state of the infrastructure. In this concern, adequate, stable, and safe equipment needs to be used. Bridges are subjected to the atmospheric conditions: wind, rain, sun irradiation, and snow. All these factors contribute to a slow but continuous interaction with the structure.

Around the bridges, several micro clima conditions may form. Some parts are directly and continuously exposed to rain such as the curbs. Other parts of the infrastructure are more indirectly exposed such as the lower and the lateral parts that are partially sheltered by the curbs (**Figure 3** left). The bridge girders (**Figure 1** center-left) are almost completely protected by the structure itself by the direct interaction with the weather parameters such as rain and the direct sun irradiation, while the lower section of the piles is occasionally directly exposed.

In this concern, the type and the extent of the degradation may be significantly different. Generally, the South side of the bridges is subjected to the cyclic exposure to rain and sun. The relatively higher presence of sun irradiation lowers the frequent presence humidity. On the other hand, the cyclic temperature variation and the subsequent relative humidity variation cause an increased carbonation of the side elements, partially covered by the curbs (**Figure 3** left). The high penetration depth of the carbonation decreases the pH down to the rebar and causes the rebar corrosion. This results in a clear and visible cracking located along the vertical rebars (**Figure 3** center). The spalling of the concrete cover and the resulting direct contact of the rebars to the atmosphere pushes the corrosion to an advanced stage. The corrosion products detach from the rebars. The steel rebars may start to exhibit a diameter loss (**Figure 3** right).

In the North side, the curbs are constantly exposed to rainfall. and only an indirect sun irradiation is present. Therefore, a more constant presence of humidity and a growth of organic deposits is seen. Consequently, the depth of carbonation, even after decades, can be until 10 orders of magnitude lower as compared to the South side (**Figure 4** left).

Generally, it can be stated that no direct constant correlation exists between the concrete cover and the carbonation depth. As stated earlier, the concrete quality and type of cyclic exposition play a relevant role. Shoulders, piles, beams, and shelves exhibit a similar carbonation depth, while the struts and especially the curbs show

Figure 3.
Corrosion induced by carbonation along the lateral parts of an arch bridge in service for over 40 years. Partial direct atmospheric sheltering (left), concrete spalling (center), and rebar diameter reduction (right).

Figure 4.
Low carbonation (left) of the north side curb (center-left) detected with the phenolphthalein method directly on-site [19] and lower elements of an arch bridge with the presence of carbonation, humidity, and water infiltration (center-right) from the upper parts of the infrastructure and corrosion (right).

the lowest carbonation. This latter behavior is due to the direct atmospheric rainfall and splashing exposition of these elements. It must also be said that the carbonation conditions and the direct exposition to the atmospheric elements may vary the extent of the degradation and frequent windy conditions, and cold climate may largely influence the type and extent of the degradation due to carbonation.

The compressive strength is related to the total porosity of concrete. This latter parameter influences the entrance of pollutants within the cementitious materials. Hence, a high porosity allows more easily the CO_2 to enter within the microstructure, although the presence of a porosity and a capillary porosity up to the surface of the concrete are not the only factors that affect the carbonation depth but also as previously stated. In fact, during that time, the aging of the infrastructure causes a general trend toward slightly higher compressive strength, but also a tendency to form microcracks. Consequently, within a compressive range between 40 and 120 Mpa, we may find total porosities variation from 10 to 20%. This is a quite wide range. The carbonation is only indirectly correlated with the total porosity because the capillary pores play a more significant role in the CO_2 penetration. The shoulders may exhibit a relevant variation in the carbonation for strengths up to 65 MPa. This variation is seen for in service concrete bridge over the decades within compressive strength range from 45 to 65 MPa. On the other hand, at higher strengths, the relationship between strength and mean carbonation tends to be more evident [20]. On the contrary, a too high rigidity of the material may enhance the formation of cracks and in turn, increase the penetration of CO_2 and the carbonation.

The splashing of chloride-rich waters and the infiltration from the upper to the lower parts on an infrastructure are the main concern for the rebar localized corrosion. The shoulders, heads, and curbs are subjected to these contaminations and often exhibit high chloride content within the concrete with values up to 0.400% referred to the concrete mass. The struts, piles, and beams generally exhibit lower values. As for the carbonation, the many parts of a bridge show different chloride contamination. Shoulders and heads may exhibit lower concentration with depth due to the washing effects. Although a high concentration is found at the surface. A limit of 0.025% by concrete mass is often set to initiate corrosion processes along the rebars [21]. Nevertheless, many laboratories test set-ups are not realistic enough to determine the threshold values and the corrosion behavior in the field [22]. Therefore, critical chloride content data were gathered from real infrastructures [23]. In addition, the exposition to chloride splashing and the alternate wetting and drying cycles may partially evaporate the water and lead to intermittent increased chloride concentrations well above the theoretical limits. This may happen on the surface several times during the service life of bridge elements. These cyclic processes and the combined actions between carbonation, water leaching, freeze, and thaw, as well as alternate chloride penetration steps and salt crystallization, may accelerate the degradation [9]. Chloride penetration and carbonation are some of the main adverse actions that can occur in reinforced concrete. It is also important to know that in service bridges elements, such as shelves, shoulders, and curbs, exist a low correlation between a low mean carbonation and a high chloride content. The shelves and the shoulders are exposed to wetting/drying cycles. Therefore, carbonation and chloride contamination may be both present. The curbs are highly exposed to chloride-enriched water splashing, and the chloride content is high down to 40 mm depth. This may also be caused by the mechanical leaching action, the water droplet, and the enhancement of the surface porosity. On the contrary, struts, beams, and piles are sheltered from rainfall or splashing and the corrosion induced by carbonation takes the lead as a main degrading action [24]. In very cold climas, usually outside the European alpine regions, the freeze/thaw and the alkali-aggregate damage may often be considered as main deterioration mechanisms [9].

5. The durability of the components

The environmental exposition can range in different time slots: yearly, seasonally, monthly, weekly, daily, and within hours. Rapid and relevant short-time atmospheric changes are particularly seen along the alpine region, starting from 1000 meters above sea level. The cyclic exposure conditions are often partially disregarded with respect to the durability. Consequently, to ease the interpretation of the environmental interactions, such as carbonation, chloride migration, freeze, and thaw resistance, these long-term degradation mechanisms are separately tested within the laboratories [3, 21]. In this manner, the link with the real field behavior is often incomplete with respect to the onsite durability of bridges.

The different elements of bridges are differently exposed to the atmospheric conditions [25]. Therefore, the carbonation and the chloride penetration may vary. Additional investigations on over 35 years old highway and road bridges indicate a variable extent of the carbonation. The shelves exhibit changing values from very low up to 20 mm (**Figure 5** upper row left) and the beams a slightly higher trend (**Figure 5** upper row center). The curbs indicate the lowest values among all the bridge elements, with values that only locally reach 10 mm (**Figure 5** upper row right). In most of the cases, the carbonation of these elements remains well below 5 mm during the decades, especially along the North-East exposed bridge sides, where the humidity is more often present. These elements are directly exposed to rainfall and the atmospheric condition, but above all, they are continuously splashed by the traffic vehicles during the rainy days. The presence of water and humidity largely hinders the CO_2 penetration, necessary to cause the carbonation process. The decks also exhibit relatively low values (**Figure 5** lower row left). This may be accounted for the protection of these elements by the polymer-based coating water sealing material and mastic asphalts, which also serve as a partial CO_2 barrier. On the contrary, the unsheltered and direct exposition to the atmosphere of the shoulders and the piles causes an increased carbonation up to 35 mm, although with variable values depending on the bridge concrete quality and sampling location (**Figure 5** lower row, center and right). In this concern, only a slight

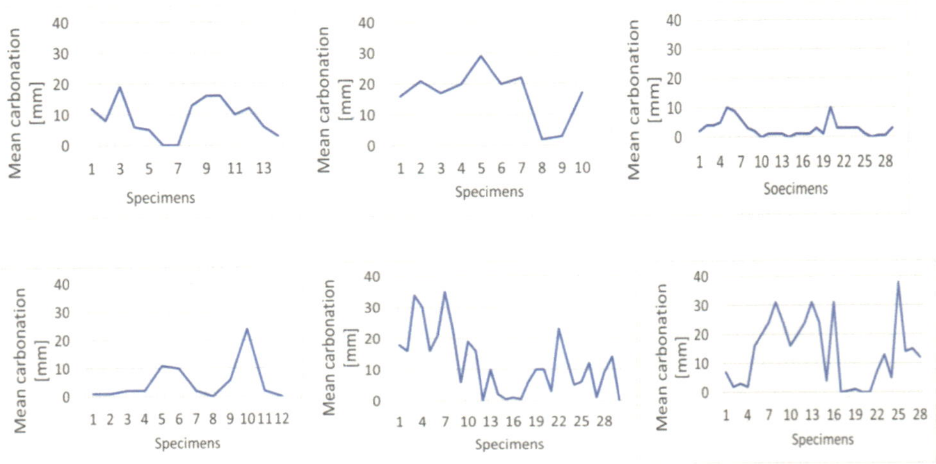

Figure 5.
Carbonation depth profile of bridge elements along a highway in service for over 40 years in a south alpine region. Upper row: (left)-shelves, (center)-beams, (right)-curbs. Lower row: (left)-decks, (center)-shoulders, (right)-piles.

correlation in the field exists between the carbonation and the compressive strength for values up to 65 MPa, depending on the bridge element considered. Only compressive strength values above 65 Mpa with a significant reduction in the porosity may significantly hinder the carbonation [20]. On the other hand, it must also be stated that the carbonation largely depends on the $Ca(OH)_2$ availability and the cyclic relative humidity ranges, particularly present in the field.

The chloride content is relatively low for the shelves (**Figure 6** upper row left) and the beams (**Figure 6** upper row center), which are partially sheltered by the direct splashing and exposition to the chloride-rich water. They generally show values below the conventional chloride critical value of 0.025% referred to the concrete mass [21]. The curbs exhibit the highest chloride content with values above 0.1% (**Figure 6** upper row right). This is due to more frequent contact of chloride-rich splashing water. The water leaching effects that may be seen in the shoulder and heads with chloride enrichments with the depth and a decrease in the surface [24] are not seen in the curbs as horizontal elements. However, high chloride content up to 0.1% down to a depth of 40 mm is quite a constant observation and was observed also in these bridges. As mentioned, the decks are protected by the water sealing coating and the mastic asphalt or asphalt concrete. Nonetheless, local chloride enrichments above 0.1% may also be intense (**Figure 6** lower row left). This may be due to local defects of the protection polymer membrane junctions or bituminous layer cracking that allows the contaminated water to penetrate and reach the concrete deck. The shoulders also belong to the bridge elements with a variable but relatively intense chloride concentration. This is often due to unsealed junctions (**Figure 6** lower row center). While the piles exhibit a lowering of the values due to the partially sheltered conditions so that the chloride-rich waters cannot often come directly in contact with these latter elements (**Figure 6** lower row right).

With depth, the chloride concentration may decrease for all elements as it can be seen for the depth 10–20 mm (**Figure** 7).

A further decrease is seen for the depth 20–30 mm (**Figure 8**). Nevertheless, chloride enrichments with depth may be partially found along support walls and underpasses due to the water leaching action along the surface [25].

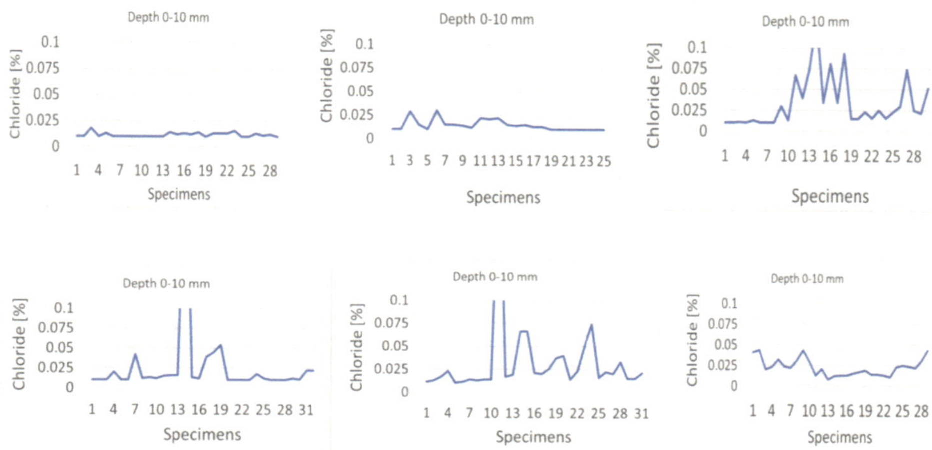

Figure 6.
Chloride depth profile of bridge elements of a highway in service for over 40 years in a south alpine region at a depth 0–10 mm. Upper row: (left)-shelves, (center)-beams, (right)-curbs. Lower row: (left)-decks, (center)-shoulders, (right)-piles.

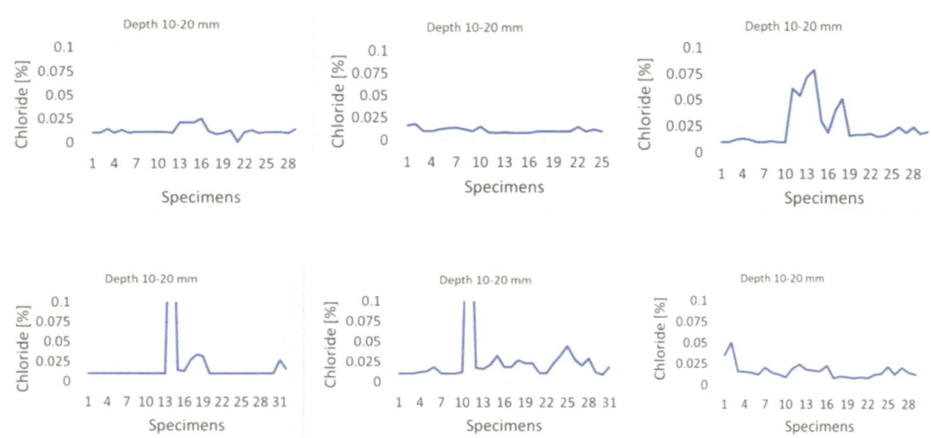

Figure 7.
Chloride depth profile of bridge elements at a depth 10–20 mm. Upper row: (left)-shelves, (center)-beams, (right)-curbs. Lower row: (left)-decks, (center)-shoulders, (right)-piles.

The carbonation and the chloride penetration are processes that lead to rebar corrosion. The former decreases the pH and causes uniform corrosion. This can be more easily monitored also from a visual point of view. The pressure arised from the formation of rebar corrosion products leads to internal tension and cracking within the concrete cover and spalling is observed. At that point, the water and the CO_2 ingress further accelerate the process. On the other hand, a similar basic tensional state is observed by chloride-induced attack of the steel rebars. However, the localized form of corrosion promoted by the chlorides leads to premature diameter reduction of the rebar. This process may enhance the speed of deterioration and lead to structural adverse issues. It must also be emphasized that in an alpine space, increased carbonation and chloride penetration are often both observed.

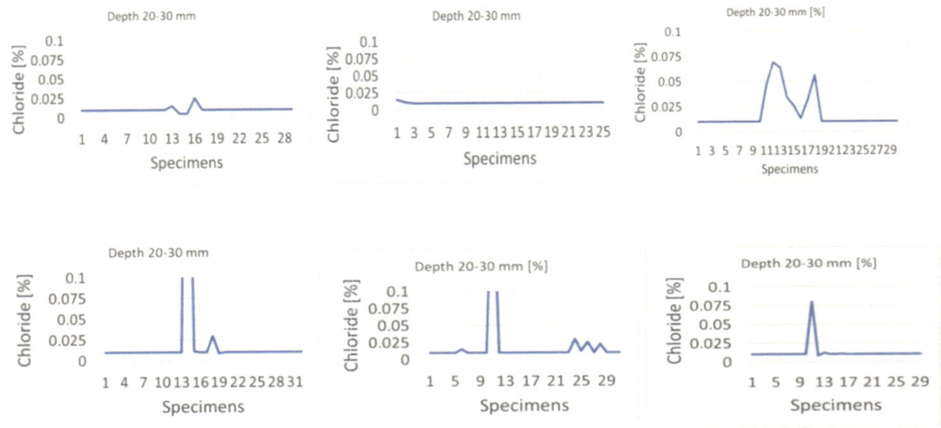

Figure 8.
Chloride depth profile of bridge elements at a depth 20–30 mm. Upper row: (left)-shelves, (center)-beams, (right)-curbs. Lower row: (left)-decks, (center)-shoulders, (right)-piles.

Consequently, the combined action of these degrading mechanisms on reinforced bridge structures decreases their durability at a faster speed as often wrongly detected by the actual laboratory tests [3, 21]. In this regard, the increased detrimental effects triggered by the environmental combined actions are higher than the single action considered separately [9, 26–29].

Hence, a new method of approaching the durability of bridges has to be developed in order to consider the system as a whole. To implement the laboratory-field relationship, it is necessary for a more realistic simulation of the field conditions. This can be achieved solely by the combination of the several factors occurring on side. On the other side, the combination of the single actions, such as carbonation, chloride penetration, freeze and thaw, wetting/drying cycles, and water leaching, must be tested in a reasonable sequence. In addition, the process cannot be accelerated above some clear defined limits to avoid distortion of the real conditions on-site. Temperature and cyclic exposure frequency must be set at an adequate level. Artificial cracking may be introduced by loading the samples. Nevertheless, firstly, it would be

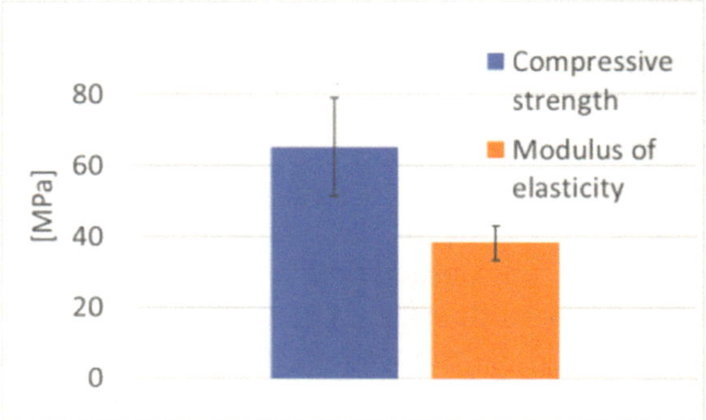

Figure 9.
Compressive strength and modulus of elasticity of the bridge elements investigated.

wise to gain a deep knowledge of the surface micro-cracks or cracks pattern present along the infrastructures. This will allow us to better simulate the load-crack-creep relation. The relatively high compressive strength often attained in real structures, and the resulting high modulus of elasticity may often promote premature cracking and reduce the durability. Modulus of elasticities up to 60 MPa was measured. These values need to be avoided in the initial cement hydration phase. A task not carefully considered in the last two decades. The premature high stiffness may be related to the low porosity needed to attain the durability test requirements of the environmental exposition classes [21]. The bridges reported in this work, in service for more than 3 decades exhibit a mean compressive strength value of 65 Mpa, and a mean modulus of elasticity below 40 GPa (**Figure 9**). Quite acceptable values with respect to the ductility and capability of the materials to partially adapt to the stresses, thus reducing the cracks and increasing the durability.

6. Junctions

The junctions are a relevant part of a bridge infrastructure. At both ends of the bridges, they have an important function in the dynamic response and dilatation behavior. On the other side, the slow but relatively frequent movement is associated with a strong solicitation of the built system in these zones. These regions may represent a weak point concerning the durability of bridges. A particular attention must be devoted to these regions. Most of the junctions are also planned in order to avoid the infiltration of pollutants and water from the above road surface. Despite the precautions, on a long-term basis, the sealing material of these junctions, such as bituminous course layers and extruded polystyrene foam XPS, are no longer able to withstand the water infiltration. Along these critical zones and beneath the sealing systems, it is necessary to foresee a water collection or conveying system that reduces the uncontrolled direct infiltration to the below reinforced concrete elements of the bridges such as the shoulders and the bridge supports (**Figure 10** left). In fact, the cracking of the bituminous sealing material (**Figure 10** left-center) and the chloride-contaminated waters from the upper parts, infiltrate along the junctions, and induce the rebar corrosion, spalling, and partial detachment of the concrete (**Figure 10** center right). The XPS material placed along some junctions (**Figure 10** right) only partially avoids a partial dirt accumulation within the gap.

Figure 10.
Junction zone along a bridge across a highway in service for over 40 years (left). Cracking (center-left), chloride-rich water infiltration (center-right), and junction filling material (right).

7. The bridge roller bearings

The bridge bearings are a fundamental part of a bridge structure and need to be panned, built, maintained, and replaced according to specific and clear guidelines. Type of bearing, service life, movement allowed, alignments, and structural, as well as seismic safety, have to be carefully evaluated [30]. This topic needs a wide discussion, but it is not the aim of the present chapter. This subchapter only describes some major issues concerning the stainless steel roller bearings. A current topic related to the durability of reinforced concrete bridges. These latter are still widely present. The planning of such roller bearings is usually done taking into account the Hertzian contact theory, although it appears that the fractures observed in some case studies indicate a partial deviation from the theory concerning some tensions arised in some failed bearings [31].

The material used for such stainless steel roller bearings is mainly martensitic stainless steel. A high-strength material with a high hardness that allows the rolling with very minimal abrasion. On the other hand, such material seems to be prone to corrosion and hydrogen embrittlement [32]. The high-strength materials, in particular steels, and high-strength steels may be particularly susceptible to surface defects, which, in turn, create initiation zone for fractures [33]. Most of the stainless steel bridge roller bearings were designed and produced during the 70's years. The rollers are in direct contact with stainless steel plates, which, in turn, are supported by large low alloyed steels S235 and S355 [34]. The stainless steel often contains a 13% weight chromium, which corresponds to a X39Cr13, 1.4031 stainless steel [35]. A different extent of degradation is seen on the rollers. Some craters may form on the surface due to humidity, contact with dirt particles, or iron powder. Aligned localized corrosion craters may be found in regions, where an infiltration of chloride-rich waters from the upper bridge junctions is seen. Corrosion may be observed when the face of the rollers remains in contact with the plates and lately, scratches or surface defects are attacked by atmospheric and pollutant agents. In this case, a sharp line of surface corrosion is seen (**Figure 11** left). Coating detachment from the lower plates can also be seen due to the movement of the roller (**Figure 11**, second from left). Corrosion and abrasion issues can be found along the central guide of the rollers (**Figure 11** center), while to reduce these phenomena gliding polymer, materials are inserted to reduce the degradation (**Figure 11** center-right). The type of damage and the extent of corrosion also depend on the service condition of the bridge, and in extreme cases, vertical cracks may be found across a roller (**Figure 11** right).

Figure 11.
Contact corrosion (left), plate coating detachment (second from left), corrosion and coating detachment along the central guide (center), Teflon sliding bands (center-right), and vertical crack (right-picture M. Paderi) on bridge roller bearings in service for over 40 years.

The detection and controlling of such damages can mainly be carried out with regular visual inspection, liquid penetration [36, 37], and ultrasound techniques [38, 39]. The former technique enables to localize surface to reach defects not visible with the bare eye, and the latter allows the identification of internal inhomogeneities, usually larger than 3 mm. This defect size may be a partial limiting factor of technique.

A constant maintenance is necessary for such types of high-strength martensitic stainless steels, regardless of their chemical composition and type or the presence of a low alloyed steels in the core of the cylinders. All surfaces need to be protected from corrosion, especially the contact with aggressive NaCl-rich waters must be avoided. Sealed junctions, adequate oil, or grease may help the protection. The alignments of the roller bearing systems must be constantly verified in order to avoid the degradation and the resulting structural-related issues. Tensions and abrasion along the central guide elements must be eliminated by appropriate alignments and by inserting sliding materials along the contact zone rollers-central metallic guide. It is also necessary to limit the accumulation of dirt and sand particles, particularly along the central zones to avoid oxygen depletion and instability of the protective oxide coating. These regions are particularly subjected to rupture, as well as the lateral cylindric parts, where fracture may arise and propagate longitudinally along the bearings [31]. It is necessary to avoid the cleaning of the rollers with abrasive papers. The abrasion is detrimental since it forms scratches on the bearing surface. High-strength metals may be particularly susceptible to surface defects. They may act as initiation sites for localized corrosion and fracture. Therefore, any type of surface anomaly must be carefully evaluated [40], by also taking into account the loading conditions.

8. The ground embedded infrastructure

The foundation piles are a part of a bridge completely embedded into the ground. The basement of pylons is also stuck into the ground. The pylons foundation piles penetrate down to variable depths. In the case presented here, 10 to 15 foundation piles were present for each highway viaduct pylon and reached a depth ranging from 10 to 15 meters. The foundation piles are made of rebar-reinforced concrete or steel cylinders filled with cementitious material and covered with concrete or special injection mortars on the external side. The two types of cementitious materials may exhibit significant differences for the internal and external applications with respect to the water permeability (from 2 $g/m^{2}*h$ to 17 $g/m^{2}*h$, respectively) and compression strength (from 70 to 170 MPa, respectively). Other types of piles are also possible. The concrete carbonation is largely lowered within the ground. Hence, the corrosion induced by carbonation is barely an issue. The chloride attack is also scarcely present unless the chloride-contaminated water from the bridge superstructure is reaching the ground due to inadequate water drainage and conveying.

The sampling and quality control of such ground embedded elements are not an easy task. During the placing but also for the inspections, safety precautions must be taken to avoid ground subsidence into the excavated places around the piles. Furthermore, during rainy days the presence of high-level groundwaters may hinder the work, and some of the water have to be pumped out of the excavation (**Figure 12** left). The groundwater level variation and flux are responsible for relevant leaching effects along the pylon basements. The water level variation can also be seen directly on foundation piles, where some slight horizontal marks of the water level can be observed. Nonetheless, the low permeability of the external covering mortars largely reduces the

Figure 12.
Safety concern for the inspection of the upper parts of foundation piles (left), pylon groundwater leaching phenomena (left-center), rebars (center-right), and steel conservation state of the piles (right). Highway bridges in service for over 40 years, still in service.

water leaching phenomena. The compressive strength class for pylons and foundation piles may be relatively high ranging from a class C 70/85 to a class C 40/50. Some pylon basements may exhibit a clear leaching due to the groundwater level change. In these cases, horizontal leaching lines are clearly visible and cause the disgregation of the cementitious binders, thus exposing the stone aggregate (**Figure 12** center-left). Consequently, the surface compressive strength class of the cementitious materials may drop to C 25/30. Groundwater flux velocity, the frequency of level change, and the chemistry parameters, such as carbonate hardness, free CO_2 content, chloride, sulfate, and pH, may significantly affect the leaching resistance.

The corrosion behavior of the buried metallic parts is largely influenced from their embedment into the cementitious materials. The cover depth may range from 30 mm to 80 mm. It is worth emphasizing the changes that may occur in the injection of materials with depth. Down to 10–15 meters, the injection conditions may vary and the cementitious cover of the steels may not be homogeneous. In addition, a change in the ground and rock composition may expose the foundation piles to different conditions. Local depletion or enrichments of oxygen may create macro-element cells that increase the corrosion of the oxygen depleted zones. This fact needs to be particularly considered for clay-rich and the progressive change to sand-rich soils must be carefully taken into account. A variable extent of the concrete leaching may be present depending on the groundwater situation, but both types of foundation piles may not exhibit a significant corrosion of the rebar or steel piles if they are embedded in an adequate quality of cement-based materials (**Figure 12** center-right; right). All parts of an infrastructure buried into the ground and with a relative extension in size may be susceptible of stray current corrosion. This may happen in the vicinity of direct current railway lines. Therefore, the electrochemical potential of the buried reinforced concrete steel rebar or metallic elements needs to be monitored [41].

9. Conclusions and future outlook

The reinforced concrete bridges belong to the most relevant infrastructures in the construction field. The mechanical performance of such structures remains a key issue. Nonetheless, the durability increased its importance with time. In fact, such structures are widely exposed to variable loading and environmental conditions. These latter largely affect the durability. The degradation of bridge elements is strongly asymmetrical. The different parts of a bridge are differently exposed to the aggressive actions. Direct and indirect exposition to the atmosphere, cyclic exposure, rain, wind, sun, freeze and thaw, chloride contamination, and carbonation are among the main degrading actions taking place around a bridge. In this concern, different microclimas arise along such infrastructures. This fact needs to be considered in the planning phase.

The higher requirements set by some concrete mix design concepts and requirements such as the water/cement ratio and the minimum cement content, as well as the laboratory tests and the normative concerning the durability, tend to decrease the general porosity of the concrete. This results in an implementation of the compressive strength of the materials, especially at early age. The modulus of elasticity and the stiffness also increase. Consequently, premature cracking may often occur in the field. This fact significantly reduces the durability. Therefore, a variable relationship between laboratory tests and real field durability is often ascertained.

In this concern, it is necessary to rethink some concepts about the concrete mix design. It is necessary to lower the clinker cement component in the cementitious binders and to increase the use of composite cement. In addition, the use of recycled concrete aggregates also needs to be considered for infrastructure by limiting the percentage of addition to reasonable values. This enables to attain the required durability. Furthermore, it is worth to implement the relationship between laboratory durability tests and real field behavior. This may be partially accomplished by exposing the laboratory samples to combined degradation actions as it is in the field, in order to obtain more reliable laboratory results and more durable bridges in the future.

Acknowledgements

The author would like to thank the collaborators and technicians of the institute of materials and construction for the field interactions, sampling, and testing of the materials.

Conflict of interest

The author declares no conflict of interest.

Author details

Christian Paglia
University of Applied Sciences of Southern Switzerland, Supsi, V. Flora Ruchat, Mendrisio, Switzerland

*Address all correspondence to: christian.paglia@supsi.ch

IntechOpen

© 2024 The Author(s). Licensee IntechOpen. This chapter is distributed under the terms of the Creative Commons Attribution License (http://creativecommons.org/licenses/by/3.0), which permits unrestricted use, distribution, and reproduction in any medium, provided the original work is properly cited.

References

[1] Normative ISO19595, Natural aggregates for concrete, 2017

[2] Normative SIA 262, Concrete Constructions, 2013

[3] Normative SIA 262/1, Concrete Constructions, Complimentary Specifications, 2013

[4] Normative EN ISO 15630-1, Rebar for Reinforced Concrete—Test Methods—Part 1: Rebar, Rolls, Threads for Reinforced Concrete, 2019

[5] Juenger MC, Siddique GR. Recent advances in understanding the role of supplementary cementitious materials in concrete. Cement and Concrete Research. 2015;**78**(Part A):71-78

[6] Paglia C, Antonietti S, Corredig G. The environmental degradation of cement-based materials due to alkali-silica reaction: Case studies. In: The Second International Conference on Construction Materials and Structures, online event. India: ICCMS; 2022

[7] Paglia C, Wombacher F, Böhni H. The influence of alkali-free and alkaline shotcrete accelerators within cement systems. Influence of the temperature on the sulphate attack mechanisms and damage, Cement and concrete research. 2003;**33**:387-395

[8] Paglia C, Mosca C. The hydrochloric acid attack of reinforced concrete in a gold refinery. In: AMPP Annual Conference. San Antonio, Texas, US: The association for materials protection and performance; 2022, online

[9] Wang R, Zhang Q, Li Y. Deterioration of concrete under the coupling effects of freeze–thaw cycles and other actions: A review. Construction and Building Materials. 2022;**319**

[10] Lea's Chemistry of Cement and Concrete, Editors: Peter Hewlett, Martin Liska, 5th Edition, 2019

[11] Papadakis VG, Vayenas CG, Fardis MN. Physical and chemical characteristics affecting the durability of concrete. ACI Materials Journal. 1991;**88**(2)

[12] Bassat MB, Nixon PJ, Hardcastle J. The effect of differences in the composition of Portland cement on the properties of hardened concrete. Magazine of Concrete Research. 1990;**42**(151):59-66

[13] Matthews JD. Carbonation of ten-year concretes with and without pulverized-fuel ash. In: Proc. Ashtech Conf. London; 1984. p. 12

[14] Neville AM. Properties of Concrete. 4th ed. London: Longman; 1999

[15] Schubert P and Wesche K, Einfluss der Karbonatisierung auf die Eigenschaften von Zementmörtel, Research report No. F 16, Institut für Bauforschung BWTH, Aachen 1974

[16] Papadakis VG, Fardis MN, Vayenas CG. Effect of composition environmental factors and cement-lime mortar coating on concrete carbonation. Materials and Structures. 1992;**25**(149)

[17] Liu Y, Weyers RE. Modeling the time-to-corrosion cracking in chloride contaminated reinforced concrete structures. ACI Materials Journal. 1998;**95**:6

[18] Ye H, Jin N, Jin X, Fu C. Model of chloride penetration into cracked concrete subject to drying–wetting

cycles. Construction and Building Materials. 2012;**36**:259-269

[19] SN EN 14630 norm, Determination of carbonation depth in hardened concrete by the phenolpthalein method, 2006

[20] Paglia C, Antonietti S, Mosca C. The relationship between compressive strength and carbonation for in service concrete elements. In: The 8th International Conference on Advanced Engineering and Technology. Incheon National University South Korea, online,; 2022

[21] Normative EN 206/1, Part 1: Concrete, Specifications, Performance, Production and Conformity, 2021

[22] Angst U, Elsener B, Larsen CK, Vennesland Ø. Critical chloride content in reinforced concrete — A review. Cement and Concrete Research. 2009;**39**:1122-1138

[23] Boschmann C, Kathler E, Angst U, Aguilar AM, Elsener B. Data article—A novel approach to systematically collect critical chloride contents in concrete in an open access data base. Data in Brief. 2019;**27**

[24] Paglia C, Antonietti S, Mosca C. The degradation of concrete bridge elements. In: 5th International Conference on Building Materials and Materials Engineering. Spain: Barcellona; 2021

[25] Paglia C, Antonietti S. The carbonation and chloride penetration along highway concrete structures in a south alpine space. In: 8th International Conference on Concrete Technology. UK: Concrete Solutions Leeds; 2022

[26] Kuosa H, Ferreira RM, Holt E, Leivo M, Vesikari E. Effect of coupled deterioration by freeze–thaw, carbonation and chlorides on concrete service life. Cement & Concrete Composites. 2014;**47**:32-40

[27] Al-Ameeri A, S, Rafiq MI, Tsioulou O. Combined impact of carbonation and crack width on the chloride penetration and corrosion resistance of concrete structures. Cement and Concrete Composites. 2021;**115**

[28] Malheiro R, Camões A, Ferreira RM, Meira G, Amorim T. Effect of Carbonation on the Chloride Diffusion of Mortar Specimens Exposed to Cyclic Wetting, XIII international conference on durability of building materials and components. Sao Paulo, Brazil; 2014

[29] Eguez AH. Durability of concrete under combined attack. frost salt scaling in carbonated concrete and chloride ingress in mechanically loaded concrete. In: Ph. Thesis, Magnel Laboratory for Concrete Research, Department of Structural Engineering, Faculty of Engineering and Architecture. Belgium: Ghent University; 2017

[30] Normative EN 1337-1, Structural Bearings—Part 1, General Design Rules, 2000.

[31] Noury P, Eriksson K. Failure analysis of martensitic stainless steel bridge roller bearings. Engineering Failure Analysis. 2017;**79**:1017-1030

[32] Chandra K, Kain V, Srinivasan N, Samajdar I, Balasubrahmanian AK. Temper embrittlement and corrosion behaviour of martensitic stainless steel 420. Advanced Materials Research. 2013;**794**:757-765

[33] Paglia C, Mosca C, Antonietti S. The failure of a reinforced concrete bridge high strength steel anchorage, ICEFA IX. In: The 9th International Conference on Engineering Failure Analysis. Shangai, online; 2022

[34] Normative EN 10025-2. Hot rolled products of structural steels. Part 2: Technical delivery conditions for non-alloy structural steels, 2019

[35] Normative EN 10088-2, Stainless Steels—Part 2: Technical Delivery Conditions for Sheet / Plate and Strip of Corrosion Resisting Steels for General Purposes, 2014

[36] Normative EN ISO 3452-1. Non-destructive testing. Penetrant testing. General principles, 2021

[37] Normative EN ISO 23277, Penetrant Testing. Acceptance levels; 2015

[38] Normative EN ISO 17640-B, Non Destructive Testing of Welds, 2017.

[39] Normative EN ISO 10228-3-4. Non-destructive testing of steel forgings, 2016

[40] Normative EN 1337-10 Structural bearings. Part 10. Inspection and maintenance, 2005

[41] Paglia C, Jornet A. The conservation state of highway foundation bridge piles. In: 15th International Conference on Recent Advances in Concrete Technology and Sustainability Issues, June 8-10. Milan: Italy; 2022

Chapter 3

Advanced Analytical Methods for Fatigue Assessment of Ancillary Systems in Highway Bridges

Khalid W. Al Shboul, Husam A. Alshareef and Hayder A. Rasheed

Abstract

Fatigue failure is a major concern for highway sign structures due to sustained wind-loading events, which have been recognized in many states. To ensure public safety, AASHTO specifies that structures should be designed for infinite life by maintaining wind-induced stress below their constant amplitude fatigue threshold (CAFT). However, existing structures that were not designed for fatigue may contain unnoticed fatigue cracks that are difficult to detect through visual inspection, which is also time-consuming. To address this issue, a simplified analytical inspection tool was developed and implemented into computer software. The tool assesses all critical components according to AASHTO specifications for fatigue and was used to examine a failed structure, which revealed a fatigue damage crack in the vertical weld of the mast-to-arm box connection at the upper chord level. In addition, a spatial interpolation technique was proposed using Isoparametric finite element shape functions to derive wind speed records for unsampled locations from actual data recorded at known locations. This provides a better understanding of the wind events that might be the driving source for fatigue failure of these flexible structures and facilitates fatigue-life prediction by generating a full range of wind loading. Overall, this chapter contributes to improving the safety and efficiency of highway sign structures by providing effective inspection tools and wind-speed interpolation techniques.

Keywords: traffic safety, fatigue damage, finite element, wind engineering, structural analysis, Rainflow analysis, damage detection, bridge engineering

1. Introduction

Full-span overhead, cantilever, and butterfly sign support structures are critical ancillary systems on highways that play a crucial role in enhancing accessibility, efficiency, and safety of traffic flow. These systems are responsible for supporting large traffic signs that provide essential information to the public and traffic users. However, these structures have been in service for 30–45 years, and are not designed to withstand fatigue damage resulting from various loading scenarios, such as natural wind gusts, vortex shedding, and truck-induced vibrations [1–3]. Fatigue damage often occurs in the connection details of these systems, which can lead to catastrophic failures, resulting in injuries, property damage, and traffic closures. As a result, the safety concerns associated with fatigue damage have prompted a comprehensive investigation to accurately simulate and calculate the extent of fatigue damage in these three ancillary systems, following the latest AASHTO standards.

1.1 Objective of chapter

This chapter aims to present an analytical inspection tool that utilizes natural wind-time histories and AASHTO fatigue life calculation procedures to estimate fatigue life consumption in critical spots of cantilevered, overhead, and double-cantilevered (Butterfly) structures. The focus will be on analytical modeling, structural assessment, and software development, with an emphasis on the challenges of wind load event approximation and the importance of accurate lifetime calculations to minimize the unreliability of fatigue life predictions.

2. Source of induced vibrations

Highway sign structures are critical components of the transportation infrastructure, providing safety and guidance to drivers on highways. These structures are slender and tall and are subject to diverse environmental and dynamic loading conditions, which can lead to vibrations and fatigue damage [4]. Identifying and comprehending the sources of induced vibrations that can cause fatigue damage in highway sign structures is crucial for their safety and longevity. Wind loads and traffic loads, are the main sources of induced vibrations that can cause fatigue damage in these structures. Wind-induced vibrations, in particular, are the primary source of fatigue damage in highway sign structures, while traffic-induced vibrations can also contribute to fatigue damage. Furthermore, galloping and vortex shedding are two important phenomena that can cause fatigue damage in these structures as well [5–7]. Galloping occurs when wind gusts cause the sign structure to oscillate in a twisting motion, potentially leading to high-cycle fatigue damage. Vortex shedding, on the other hand, results from wind flow around a structure, creating vortices that can cause the structure to transversely vibrate, leading to low-cycle fatigue damage. Thus, accounting for the effects of galloping and vortex shedding is crucial when evaluating the potential sources of induced vibrations and fatigue damage in highway sign structures. Proper design, maintenance, and inspection practices that consider these phenomena can aid in the prevention or mitigation of fatigue damage in these structures. However, this chapter will primarily focus on the fatigue damage caused by natural wind.

3. Natural wind modeling

Accurately estimating the wind loading events over the lifetime of a structure is crucial in reducing the uncertainties associated with fatigue life calculations. However, it is challenging to accurately characterize the wind loading events that a structure experiences during its service life, as this requires a stress fluctuation spectrum, which is often difficult to obtain. To overcome this challenge, the developing analytical natural wind-time histories is a reliable method for estimating wind speed fluctuations at specific locations and quantifying the number of stress cycles associated with each stress level that the structure experiences.

3.1 Spatial wind speed interpolation

The Isoparametric finite element shape functions can be used to estimate wind speed values for unsampled locations by interpolation from surrounding sampled locations. This process involves conventional finite element analysis procedures such as meshing, shape function evaluation, and solving for unknowns. The generated wind speed records, including mean and high speeds, are used to produce a synthetic daily wind time history. Rainflow analysis is then applied to convert the irregular wind speed history into a constant-amplitude cycle-loading. This produces a comprehensive database of wind speed versus a number of cycles for fatigue analysis, which is the main input for the fatigue calculation algorithm. The algorithm calculates the damage associated with each stress level resulting from each wind speed value in the database. Let us consider the state of Kansas as an example to demonstrate the procedure, with more details available in Refs. [4, 8, 9]. After discretizing the domain into the finite number of elements to utilize all the available data, the Kansas map was divided into 12 geometrical interpolation zones to cover the entire domain, combined with quadrilateral and triangular shapes, as shown in **Figure 1**.

The coordinates of the center of main cities (element corners) and central coordinates of the county within the element were obtained using ArcGIS [10] in terms of latitude and longitude. The county's coordinates could be written as a linear combination of the four corner coordinates using the Isoparametric shape functions as in Eq. (1). The shape functions were used to express the coordinates of the center of the county in terms of the coordinates of the cities in each element.

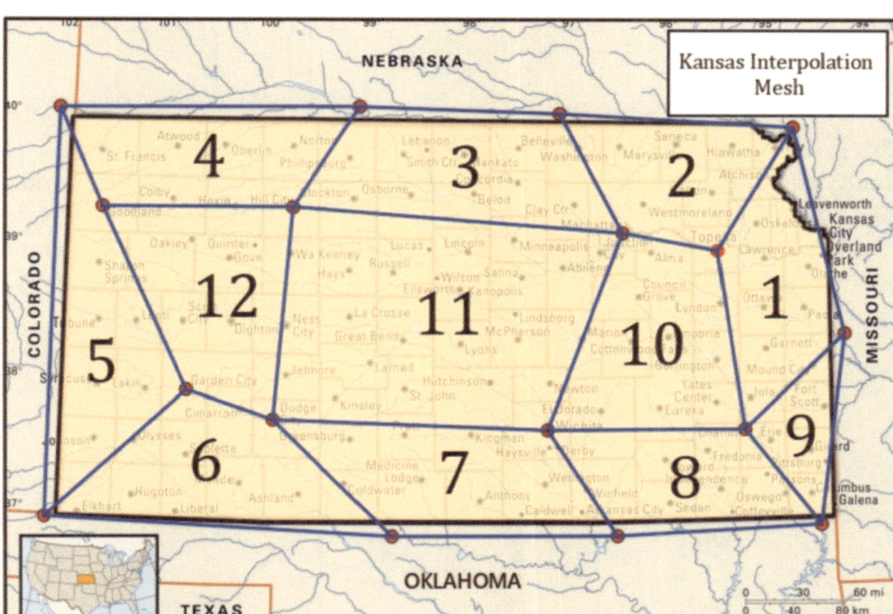

Figure 1.
Kansas interpolation zones.

$$\begin{Bmatrix} X \\ Y \end{Bmatrix} = \begin{bmatrix} X_1 & X_2 & X_3 & X_4 \\ Y_1 & Y_2 & Y_3 & Y_4 \end{bmatrix} \begin{Bmatrix} N_1 \\ N_2 \\ N_3 \\ N_4 \end{Bmatrix} \tag{1}$$

where the Ni are the FE shape functions as expressed in Eq. (2) and Eq. (3) for quadrilateral and triangular elements, respectively.

$$\begin{Bmatrix} N_1 = \frac{1}{4}(1-\xi)(1-\eta) \\ N_2 = \frac{1}{4}(1+\xi)(1-\eta) \\ N_3 = \frac{1}{4}(1+\xi)(1+\eta) \\ N_4 = \frac{1}{4}(1-r)(1+\eta) \end{Bmatrix} \tag{2}$$

$$\begin{Bmatrix} N_1 = \frac{1}{4}(1-\xi)(1-\eta) \\ N_2 = \frac{1}{4}(1+\xi)(1-\eta) \\ N_3 = \frac{1}{2}(1+\eta) \end{Bmatrix} \tag{3}$$

Using the unity property of the shape functions, Eq. (1) could be re-written as

$$\begin{Bmatrix} X \\ Y \\ 1 \end{Bmatrix} = \begin{bmatrix} X_1 & X_2 & X_3 & X_4 \\ Y_1 & Y_2 & Y_3 & Y_4 \\ 1 & 1 & 1 & 1 \end{bmatrix} \begin{Bmatrix} N_1 \\ N_2 \\ N_3 \\ N_4 \end{Bmatrix} \tag{4}$$

This is an underdetermined system of equations; we wish to derive a fourth continuity equation using the physical area of the element. The total area of a quadrilateral element is composed of a summation of the area of four triangles intersected at an arbitrary point inside the element at (X, Y), which represents the coordinates of the county as shown in **Figure 2.**

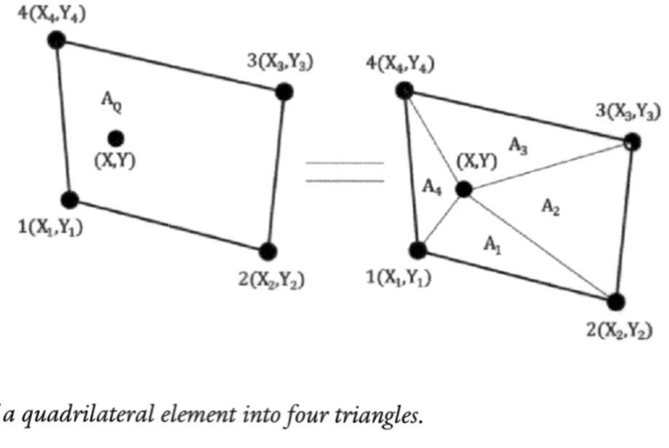

Figure 2.
Area decomposition of a quadrilateral element into four triangles.

The area of the quadrilateral could be calculated using the Gauss quadrature formula in two-dimensional regions as

$$A_Q = \int_{-1}^{1}\int_{-1}^{1} |\mathbf{J}(\xi,\eta)| d\xi d\eta = \sum_{i=1}^{2}\sum_{j=1}^{2} w_i w_j \left|\mathbf{J}\left(\xi_i,\eta_j\right)\right| \tag{5}$$

where $|\mathbf{J}(\xi,\eta)|$ is the determinant of the Jacobian matrix evaluated at the integration points $\left(\pm\frac{1}{\sqrt{3}},\pm\frac{1}{\sqrt{3}}\right)$, and $w_i = 1$, $w_j = 1$, and the Jacobian matrix is given by Eq. (6).

$$[\mathbf{J}] = \begin{bmatrix} \dfrac{\partial x}{\partial \xi} & \dfrac{\partial y}{\partial \xi} \\ \dfrac{\partial x}{\partial \eta} & \dfrac{\partial y}{\partial \eta} \end{bmatrix} = \frac{1}{4}\begin{bmatrix} -(1-\eta) & (1-\eta) & (1+\eta) & -(1+\eta) \\ -(1-\xi) & -(1+\xi) & (1+\xi) & (1-\xi) \end{bmatrix}\begin{bmatrix} x_1 & y_1 \\ x_2 & y_2 \\ x_3 & y_3 \\ x_4 & y_4 \end{bmatrix} \tag{6}$$

evaluating Eq. (5) and Eq. (6) yields

$$A_Q = \left(\frac{1}{2}x_4(y_1 - y_3) + \frac{1}{2}y_2(-y_1 + y_3) + \frac{1}{2}x_1(y_2 - y_4) + \frac{1}{2}x_3(-y_2 + y_4)\right) \tag{7}$$

At the same time, the area of a triangle is given by Eq. (8)

$$\mathbf{A_T} = 0.5 * \begin{vmatrix} 1 & x_i & y_i \\ 1 & x_j & y_j \\ 1 & x & y \end{vmatrix} \tag{8}$$

where (x_i, y_i), (x_j, y_j), the coordinates of the base and they change for each triangle while the (x, y) are the coordinates of the apex and they are constant for all the triangles. The area of all triangles could be written as:

$$\mathbf{A_1} = \frac{1}{2}\begin{vmatrix} 1 & x_1 & y_1 \\ 1 & x_2 & y_2 \\ 1 & x & y \end{vmatrix}$$

$$\mathbf{A_3} = \frac{1}{2}\begin{vmatrix} 1 & x_3 & y_3 \\ 1 & x_4 & y_4 \\ 1 & x & y \end{vmatrix}$$

$$\mathbf{A_2} = \frac{1}{2}\begin{vmatrix} 1 & x_2 & y_2 \\ 1 & x_3 & y_3 \\ 1 & x & y \end{vmatrix}$$

$$\mathbf{A_4} = \frac{1}{2}\begin{vmatrix} 1 & x_4 & y_4 \\ 1 & x_1 & y_1 \\ 1 & x & y \end{vmatrix} \tag{9}$$

Replacing the x in A_1 by

$$x = \sum_{i=1}^{4} x_i N_i \tag{10}$$

Then we have

$$A_Q - A_2 - A_3 - A_4 = \frac{1}{2}\begin{vmatrix} 1 & x_1 & y_1 \\ 1 & x_2 & y_2 \\ 1 & \sum_{i=1}^{4} x_i N_i & y \end{vmatrix} \tag{11}$$

After substituting Eq. (9) and Eq. (10) into Eq. (11) and rearranging terms, we get

$$m = AN_1 + BN_2 + CN_3 + DN_4 \tag{12}$$

where:

$$m = \frac{1}{2}(xy_1 - xy_2)$$

$$A = \frac{1}{2}x_1(y_1 - y_2)$$

$$B = \frac{1}{2}x_2(y_1 - y_2)$$

$$C = \frac{1}{2}x_3(y_1 - y_2)$$

$$D = \frac{1}{2}x_4(y_1 - y_2)$$

Then the Eq. (4) could be re-written as

$$\begin{Bmatrix} X \\ Y \\ 1 \\ \frac{1}{2}(xy_1 - xy_2) \end{Bmatrix} = \overbrace{\begin{bmatrix} x_1 & x_2 & x_3 & x_4 \\ y_1 & y_2 & y_3 & y_4 \\ 1 & 1 & 1 & 1 \\ \frac{1}{2}x_1(y_1 - y_2) & \frac{1}{2}x_2(y_1 - y_2) & \frac{1}{2}x_3(y_1 - y_2) & \frac{1}{2}x_4(y_1 - y_2) \end{bmatrix}}^{\mathbf{A}} \begin{Bmatrix} N_1 \\ N_2 \\ N_3 \\ N_4 \end{Bmatrix} \tag{13}$$

Eq. (13) cannot be solved directly since it produces a singular matrix. However, the Moore-Penrose [11] inverse A^+ could be calculated easily for this square matrix. Eq. (13) is written in compact form as

$$\boldsymbol{b} = \mathbf{A}x \tag{14}$$

The Pseudo-inverse of the singular matrix **A** is:

$$\mathbf{A}^+ = \left(\mathbf{A}^{\mathrm{T}}\mathbf{A}\right)^{-1}\mathbf{A}^{\mathrm{T}} \tag{15}$$

And the solution of Eq. (13) will therefore be approximately obtained as.

$$\boldsymbol{x} = \mathbf{A}^{+}\boldsymbol{b} \tag{16}$$

The solution **x**, in this case, is not exact. Instead, it minimizes the quantity $\left\|\vec{b} - A\vec{x}\right\|$.By knowing the county center coordinates X and Y and the nodal cities' coordinates (x_1, y_1), (x_2, y_2), (x_3, y_3), and (x_4, y_4), the shape functions (N_1, N_2, N_3, N_4) could be calculated by solving Eq. (16). After calculating the shape functions $\{N_i\}$, medium and high wind speeds were interpolated using the nodal values for the element surrounding the county as follows:

$$\text{HWS} = N_1 * HWS_1 + N_2 * HWS_2 + N_3 * HWS_3 + N_4 * HWS_4 \tag{17}$$

$$\text{MWS} = N_1 * MWS_1 + N_2 * MWS_2 + N_3 * MWS_3 + N_4 * MWS_4 \tag{18}$$

where HWS and MWS are the interpolated high and mean wind speeds for each given day in a county, HWS_i and MWS_i are the actual high and mean wind speeds for the same day for the corner city of the element considered. It is worth mentioning that this work is not intended to produce wind-speed surfaces. Instead, it produces discrete wind speed records at the center of each county. For the triangular elements, the county coordinates could be obtained easily by inverting the coordinates matrix and multiplying it by the county coordinates vector.

3.2 Synthetic wind-time histories

The spatial and temporal variation of wind velocity has two components: a daily mean component $U(z)$ and daily fluctuating component $u\ (z, t)$, expressed through $U\ (z, t) = U(z) + u\ (z, t)$, where $U\ (z, t)$ is the varying wind speed profile during the day [12]. Because wind is a random process with dynamic behavior that cannot be entirely predicted, several researchers examined different techniques like the autoregressive (AR) model, or the real wind accord to simulate the wind fluctuation with good accuracy [13, 14]. However, for the purpose of this algorithm, it is more suitable to use the well-established Kaimal spectrum [15] to simulate the power spectral density. Eq. (19) and the weighted amplitude wave superposition represented by Eq. (20) [16] was used to generate daily time history for the entire time span.

$$S_K(f) = \frac{200U_*^2 z}{U_z\left(1 + 50\frac{fz}{U_z}\right)^{5/3}} \tag{19}$$

where S_K is the Kaimal spectrum, z is the height above the ground (10 m (33 ft.)), U_* is shear velocity, U_z is the mean wind velocity at z, and f is the specified frequency.

$$u(t) = \sum_{i=1}^{N} \sqrt{2S_if_i\Delta f}\,.\cos\left(2\pi f_i t + \varnothing_i\right) \tag{20}$$

In Eq. (20), a sensitivity calculation was performed in which 798, 80, and 40 cosine waves were used to build synthetic wind speed histories for the city of Wichita over a 45-year period and extract the number of wind cycles corresponding to each

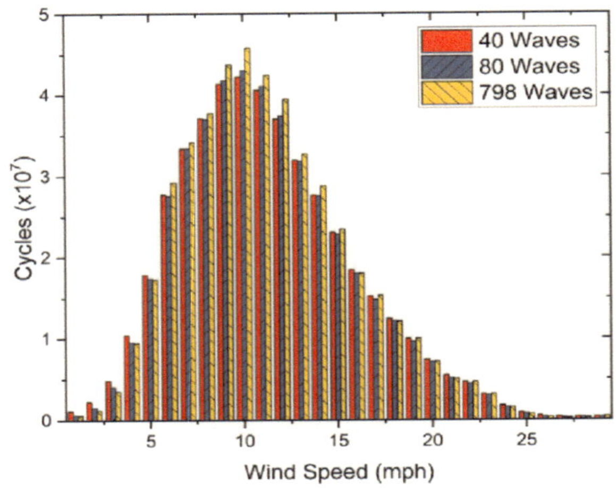

Figure 3.
Speed vs. number of cycles for 45 years in Wichita for different cosine-wave sets.

Parameter	Value
Surface roughness class	Open terrain (k = 0.005)
Height above ground	33 ft.
U_z	Vary
U_{max}	Vary
Fluctuation wind speed spectrum	Kaimal
Length of time history	One day
Time step	1 s
Frequency range	3–300 Hz
Number of cosine waves in superposition	80

Table 1.
Main parameters in wind-speed simulation.

speed. The Rainflow method was used to establish the distribution of speed versus the number of cycles for the three cosine-wave sets (**Figure 3**). The figure shows that the overall distribution was almost identical, and the cycle variation followed a Gaussian distribution. Discretization using 80 waves was an excellent trade-off between computational speed and accuracy of results to generate the 45-year wind database. More details are in Ref. [1]. **Table 1** summarizes the main parameters used in the final wind-speed simulation. **Figure 4** shows a sample of generated wind-time histories for various mean wind speeds.

The complete procedures for generating single daily time history are shown in **Figure 5** and were implemented in C# code to produce a 45-year database of wind-time histories and daily synthetic wind profiles for all counties in Kansas. After generating the database, the Rainflow counting technique [17, 18] described in detail in Section 5 was implemented to convert the irregular wind-time histories into a usable number of constant amplitude cycles.

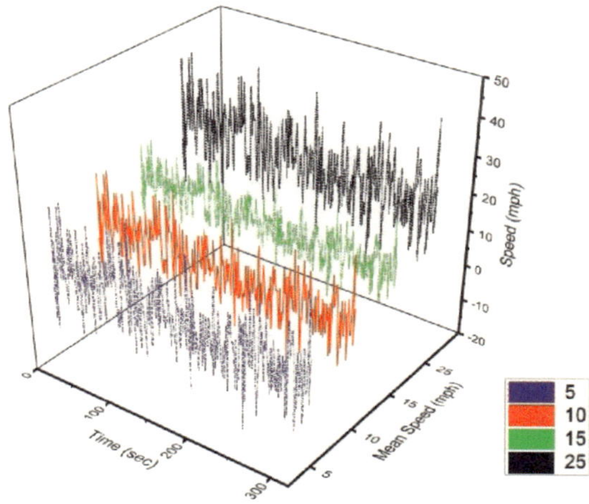

Figure 4.
Wind-time histories for various mean wind speeds.

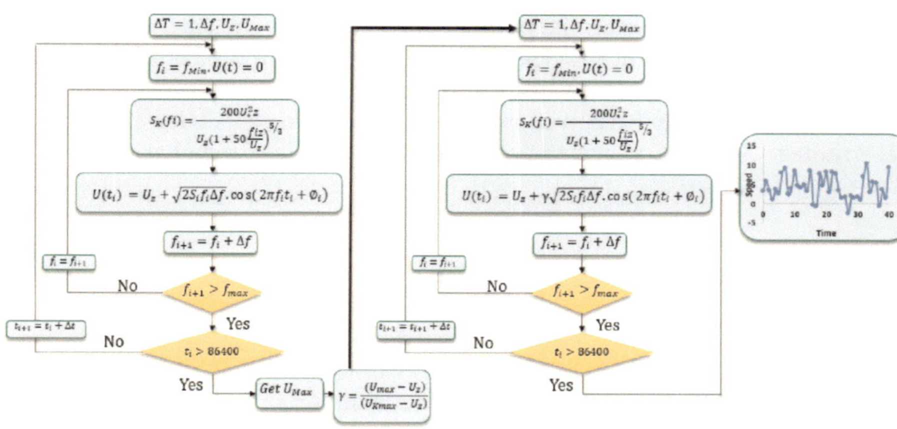

Figure 5.
Flowchart for wind-time histories generation.

4. Structural modeling

The finite element software Staad Pro V8i SS6 [19] was chosen to model the flexible sign structures and execute the static analysis. The structures were modeled using a combination of beam elements, truss elements, and plate elements with appropriate cross-sectional dimensions. The upper and lower main chords were modeled using a 2-node frame element, and they ran continuously. At the same time, the secondary members were connected at the intersection nodes and released to act as a pin connection. Moreover, the main truss chords are connected to the plates that make up both types of connections, old and new. The base support is simulated as a fixed support. In addition to that, the material used is steel with an elastic modulus of 200 GPa (29,000 ksi) for the cantilevered structures.

4.1 Mast-to-arm connection

The gusseted box connection was used in some structural models, while the ring-stiffened connection was used for other models. A 4-noded plate elements were used to model all the plates in the connection and attached directly to the main upper and lower chords at the center node. The steel plates have the same thickness as the actual structures and are connected to the pole through another plate with the same fillet weld thickness and properties. The pole was divided into sub-elements, and the plate nodes merged with pole nodes, as shown in **Figure 6**. A convergence study was conducted to verify that the model converged and to find out the mesh size that provides a mesh-independent solution for the critical stress.

4.2 Sign(s) modeling

The highway sign structures carry the sign(s) on the truss, and the sign placement varies with location, length, and height of the sign and the number of the total signs. The sign is modeled in the structure by creating a frame attached to the truss using four elements, and the sign ribs are distributed within the frame depending on the number and spacing of the ribs. The wind load was calculated based on the size of the sign then distributed equally over the ribs. Since the sign location is sensitive to the cantilever structure, the frame and the ribs are created on the structure truss at their respective location in the actual structure, **Figure 7** shows the geometry and details of the overhead structures [20].

4.3 Dynamic amplification factor (DAF)

This study utilized the static solution for certain analytical models. Due to the dynamic nature of the wind, the calculated stresses were amplified using an overall blanket average (DAF). The analytical wind modeling was carried out over a range of

(a)

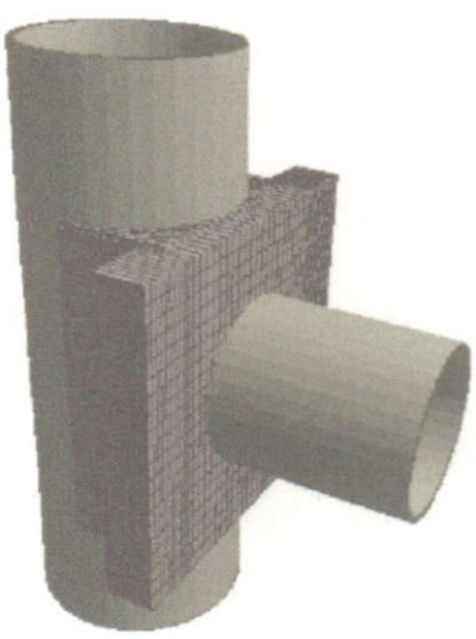
(b)

Figure 6.
(a) post-to-chord connection (b) The model.

frequencies [3–300 HZ], assuming harmonic excitations the DAF was calculated by averaging the frequency-response curve, shown in **Figure 8** for this particular range of frequencies as in Eq. (21).

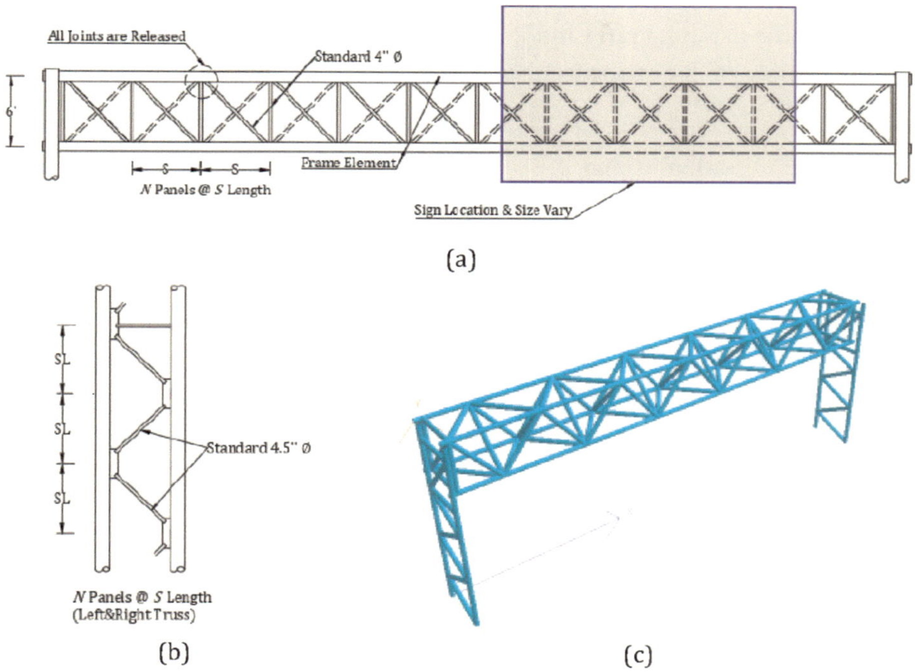

Figure 7.
(a and b) Geometry and details of the overhead truss model (c) Model of four-chord box truss.

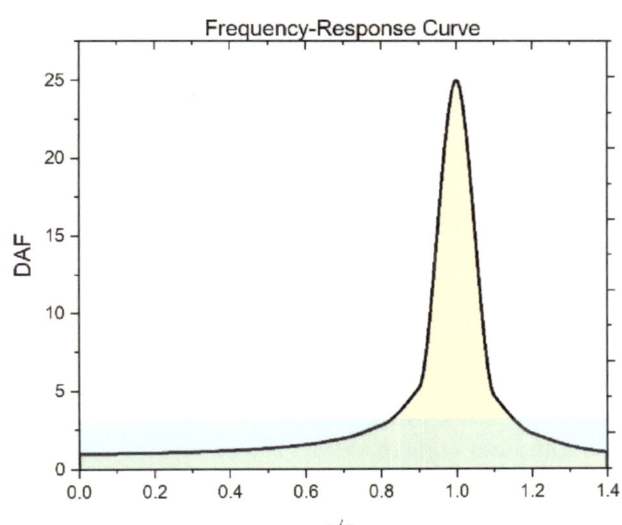

Figure 8.
Frequency-response curve and average DAF ($\xi = 0.02$).

$$DAF = \frac{\int_{0}^{1.4} \frac{dR}{\sqrt{\left(1-R^2\right)^2 + (2\xi R)^2}}}{1.4} \tag{21}$$

where ξ is the damping ratio and R = $\frac{\omega}{\omega_n}$, ω : the excitation frequency, ω_n : natural frequency of the strucutre.

4.4 Wind loading calculations

The wind loading resulting from certain wind speed was evaluated by calculating the wind pressure using AASHTO 2015 [19], Eq. (22).

$$Pz = 0.00256 K_z G V^2 I_r C_d \ (psf) \tag{22}$$

where K_z is the height and exposure factor calculated based on the height of the member and conservatively taken not to be less than 1.0, if the structure is located on a bridge, this value is taken to be 1.3. G is the gust factor = 1.14, V is the applied wind velocity (mph), I_r is the importance factor = 1.0. The drag coefficient (C_d) was considered based on the object size and shape. For the truss members, the value of C_d was taken to be 1.2, while for the signs, the value of C_d was determined based on the aspect ratio. After generating the pressure resulting from each wind speed, the pressure is multiplied by the surface area where it is applied to generate wind force.

5. The concept of fatigue and cycles counting

Metal fatigue is a phenomenon where a structure or its component undergoes damage or failure prematurely due to repetitive loading. To assess and predict the extent of fatigue damage, various analysis methods have been developed. The AASHTO Specification, for instance, uses the stress-life method to estimate fatigue lives of full-span highway sign support structures. A critical tool in this method is the S-N diagram, which plots alternating stress range versus cycles to failure. These diagrams are developed using experimental fatigue test data for different materials and structural components [21].

To generate the wind-time histories for the 45 years of data, an irregular variation of speed with time was observed. The Rainflow counting technique, developed by Matsuishi and Endo [18], was adapted to extract each wind speed cycle. This technique identifies closed hysteresis loops in a non-periodic stress response and converts irregular time histories to cycles. The algorithm borrowed from ASTM E1049 [17] was implemented into a computer code to extract the cycle database for 45 years. By implementing the Rainflow counting technique for each daily wind-time history, a speed-cycle matrix was generated that represented the number of cycles for each wind speed in a day, grouped in 0.5 mph range scale. This approach helps in predicting the extent of fatigue damage caused by wind loading in a structure, **Figure 9** shows an example of the Rainflow analysis.

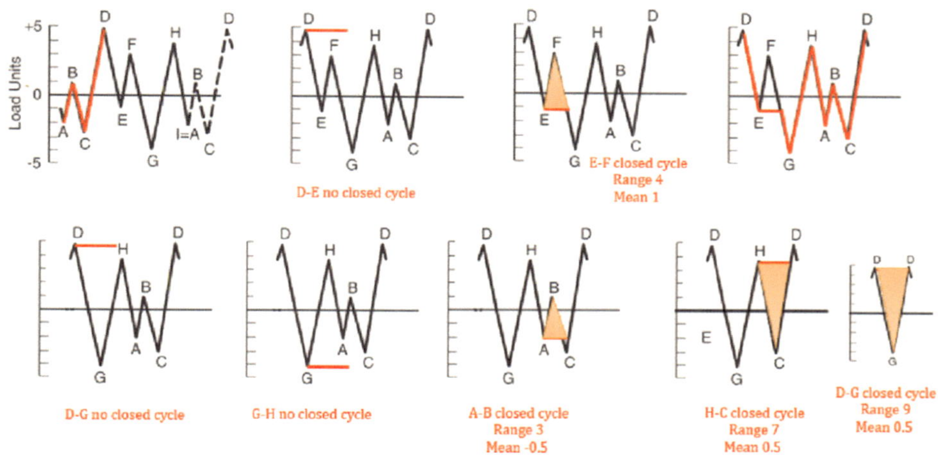

Figure 9.
Rainflow counting example.

6. Stress analysis and damage calculation

6.1 AASHTO S-N curves

AASHTO manual [22] provides *S-N* curves for different connection types based on a wide range of laboratory fatigue tests of full-scale structures. Eq. (23) could express the number of cycles to failure, where N_i is the number of cycles to failure at *i-th* stress range,$\Delta\sigma_i$is the member stress value corresponding to a wind speed value. A' is a constant associated with the component provided in AASHTO manual [22].

$$N_i = \frac{A'}{\Delta\sigma_i^3} \tag{23}$$

Table 2 describes the *S-N* equation for different components that were used in this study. Each steel *S-N* curve has a flat plateau described as the threshold in the table.

Description	A × 10^8 (ksi^3)	Threshold (ksi)	Example	Location
Plain Material	250	24	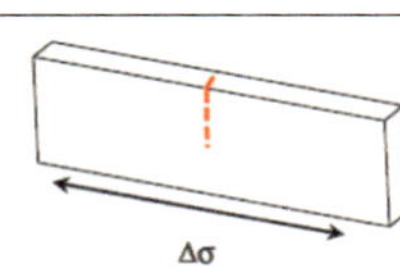	Pole Members And truss main mem.
Gusseted box connections	Infinite life	—		Mast arm connection
Ring-Stiffened box connections	Infinite life	—		Mast arm connection

Description	A × 10^8 (ksi^3)	Threshold (ksi)	Example	Location
Angle-to-gusset connections	3.9	2.6	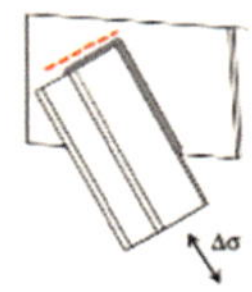	Secondary members Weld Connection
Fillet-welded tube-to transverse-plate connections	3.9	2.6	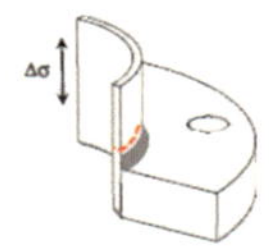	Base plate
Anchor bolts	22	7	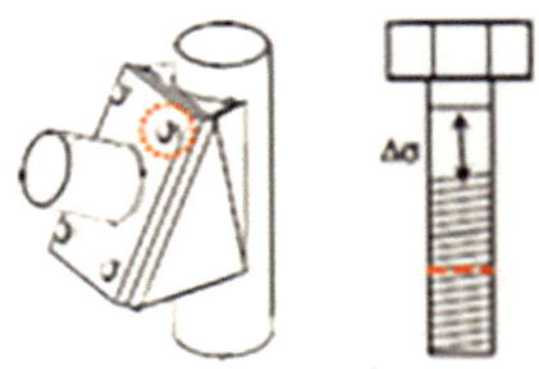	Connection bolts
Partial-penetration groove-welded mast-arm-to column pass-through connections	11	4.5	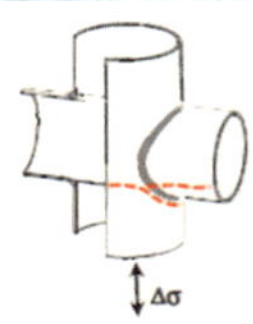	Arm weld

Table 2.
S-N equations for different structure components used by AASHTO.

Below this value, stress is assumed to have no contribution to the cumulative damage and the components have infinite life [23].

6.2 Fatigue damage and miner rule

Once the structural analysis is completed for all the wind speeds in the time period where the structure is investigated, the member end forces resulting from the FE solution were collected for each structural component associated with a wind speed. The axial stress was calculated in each critical component, as indicated in **Table 3**. The resulting stress was amplified using the DAF. Then, the calculated axial stress in each critical member is associated with an appropriate S-N curve from the AASHTO to calculate the needed number of cycles to failure. Moreover, the fractional damage was

Location	Example	Stress
Main members	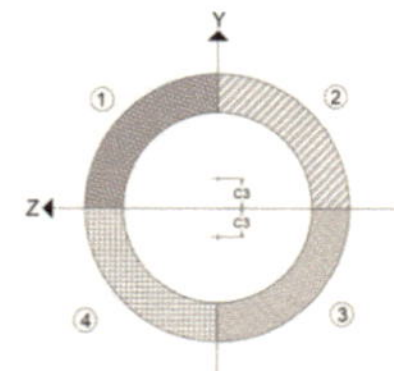	$\sigma_{x1} = \frac{My}{2C_3\frac{\pi}{2}(r_o{}^2-r_i{}^2)} - \frac{Mz}{2C_3\frac{\pi}{2}(r_o{}^2-r_i{}^2)}$ $\sigma_{x2} = \frac{-My}{2C_3\frac{\pi}{2}(r_o{}^2-r_i{}^2)} - \frac{Mz}{2C_3\frac{\pi}{2}(r_o{}^2-r_i{}^2)}$ $\sigma_{x3} = \frac{-My}{2C_3\frac{\pi}{2}(r_o{}^2-r_i{}^2)} + \frac{Mz}{2C_3\frac{\pi}{2}(r_o{}^2-r_i{}^2)}$ $\sigma_{x4} = \frac{My}{2C_3\frac{\pi}{2}(r_o{}^2-r_i{}^2)} + \frac{Mz}{2C_3\frac{\pi}{2}(r_o{}^2-r_i{}^2)}$

Location	Example	Stress
Pole	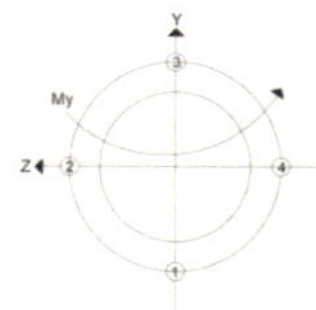	$\sigma_x = \frac{My.r}{I} + \frac{Fx}{A}$
Arm weld	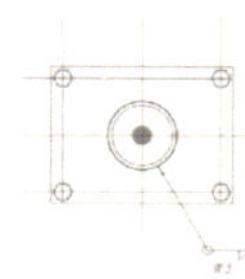	$\sigma_x = \frac{\pm My}{2C_3\frac{\pi}{2}({r_o}^2 - {r_i}^2)} \pm \frac{Mz}{2C_3\frac{\pi}{2}({r_o}^2 - {r_i}^2)}$ $\sigma_w = \sigma_x \frac{t_m}{t_{w3}}$
Connection Bolts	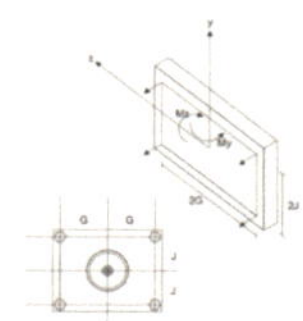	$\sigma_x = \frac{My}{4G\frac{\pi}{4}{d_b}^2} + \frac{Mz}{4J\frac{\pi}{4}{d_b}^2} + \frac{T}{4\frac{\pi}{4}{d_b}^2}$
Secondary members	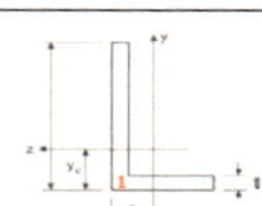	$\sigma_{x1} = -y\frac{M_z I_{yy} + M_y I_{yz}}{I_{yy} I_{zz} - I_{yz}^2} + z\frac{M_y I_{zz} + M_z I_{yz}}{I_{yy} I_{zz} - I_{yz}^2} + \frac{Fx}{A}$
Base plate weld	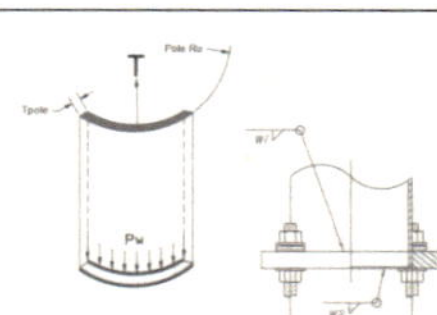	$\sigma_x = \frac{\pm My}{2C_3\frac{\pi}{2}({r_o}^2 - {r_i}^2)} \pm \frac{Mz}{2C_3\frac{\pi}{2}({r_o}^2 - {r_i}^2)}$ $\sigma_w = \sigma_x \frac{t_p}{t_{w1} + t_{w2}}$
Connection Weld	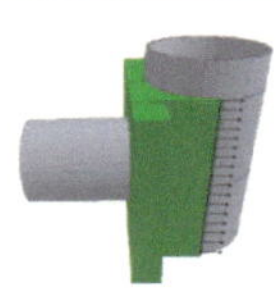	$\sigma_w = S_x \frac{t_p}{t_w}$

Table 3.
Stress calculation in different structure spots.

calculated by using Miner rule Eq. (24) to estimate the damage consumption by finding the ratio of the number. of stress cycles experienced by the structure to the number of cycles required for failure. As indicated by the *S-N* curve, only the stresses greater than the threshold were assumed to contribute to the damage.

$$D_i = \frac{n_i}{N_i} \tag{24}$$

where D_i is the damage in a specific member at a particular stress range, n_i is the number of cycles at *i-th* stress range, obtained from Rainflow analysis, N_i is the number of cycles to failure at the same stress range obtained from the S-N curve. In the scope of this work, wind time histories were generated for the 45 years of data [1], these histories represent highly irregular variations of speed with time. Rainflow counting technique [18] was used to convert the irregular time histories to cycles.

$$D = \sum_i D_i \tag{25}$$

The cumulative damage was determined by adding all the fractional damages associated with each wind speed using Eq. (25).

6.3 Case study: Cantilever structure

A cantilever highway sign structure that has been in service for 32 years was selected to test the validity of the developed approach. The structure is located in Sedgwick County, Kansas, over northbound I-235 at ramp to West Street. The structure consisted of three panels spaced at 7 ft. (2133.6 mm) and supported over a single tapered pole that has a total height of 27 ft. (8229.6 mm) and base outer diameter OD of 16 in. (406.4 mm). The main truss has design model 1 properties and consisted of multiple angle sections $3'' \times 3'' \times \frac{3}{8}''$connected by welded angle-to-gusset connections. The full geometric details are shown **Figure 10** and the general structural properties are presented in **Table 4**.

On October 30, 2019, Kansas Department of Transportation (KDOT) had performed a comprehensive field inspection to assess the condition of all the structure components as part of their regular inspection plans. The visual inspection revealed different corrosion levels for the entire structure, 50% corrosion staining was observed on the anchor bolts hardware, and full corrosion staining on the full height of the pole. In addition to that, the connection plates have corrosion staining reflected through corrosion bleed-out emanating from the weld copings. A complete fatigue crack in the vertical weld of mast-to-arm box connection at one side of the upper chord level was also observed. Close-up view of the vertical column-to-mast arm connection is shown in **Figure 11**. Obviously, the crack occurred in the entire weld toe resulting in a complete separation of the vertical splice plate from the column.

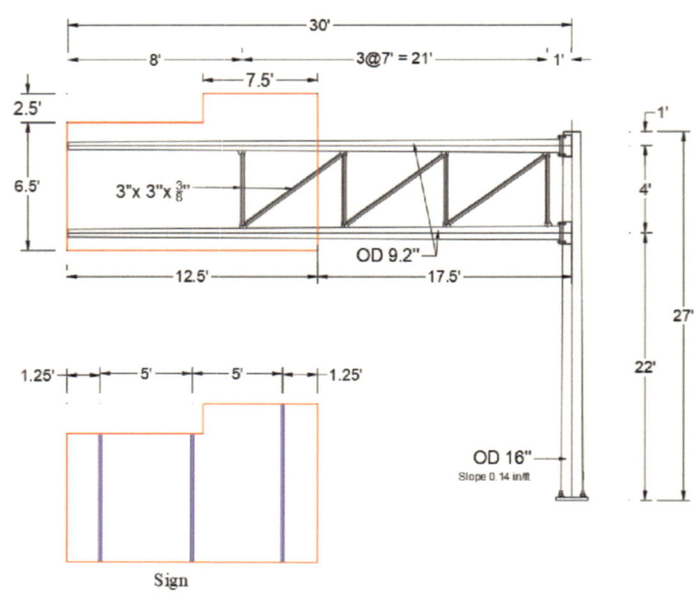

Figure 10.
Sedgwick structure geometry (all dimensions are in feet/inches).

Structural data		Original project data		Sign and attachment	
Structure type	**Cantilever**	**Date let**	**1987**	**height (ft.)**	**length (ft.)**
material	Steel	Inspection Date	2019	6.5	12.5
Arm truss span	30 ft.			2.5	7.5
Vertical clearance	18 ft.				

Table 4.
Sedgwick structure information.

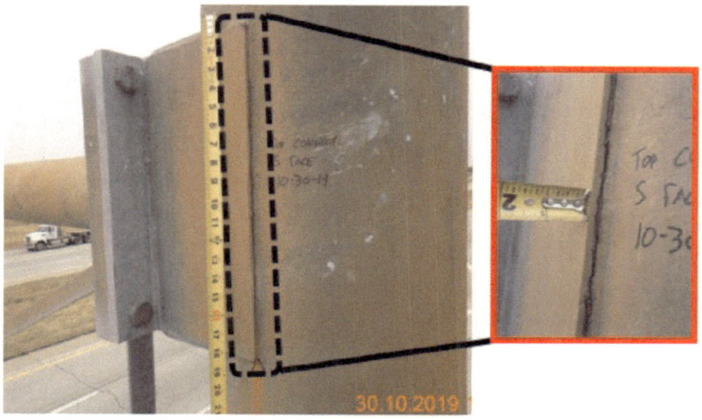

Figure 11.
Crack in the weld toe in Sedgwick structure, courtesy of KDOT. (Bureau of structural and geotechnical services).

AASHTO 2015 Structural Supports for Highway Sign LTS specifies an infinite life for both mast-arm-to-pole connections, namely, the gusseted box connection, and the ring-stiffened box if they were detailed as per the AASHTO 2015 recommendations. These connections were tested experimentally in full size and they did not develop any fatigue cracking under both in-plane and out-of-plane loading scenarios. However, in all the tested specimens, the fatigue cracking occurred in other critical locations such as the tube-to-transverse-plate welds in the mast arm, the pole, and/or hand holes [22, 24]. The connection between the side plate and the pole falls under the category of E` details in AASHTO specification having a CAFT of 2.6 ksi (18 MPa). The wind loading event for the whole structure's service life revealed a range of wind speeds (1–33 mph, 1.6–53 km/h) with the corresponding number of cycles that the structure might experience. After providing the software with the necessary information, including an approximated average corrosion reduction factor of about 17%, which reflects the existing conditions of this structure, the software starts to build and run the analysis. Upon performing the required successive analyses by STAAD pro, the software read back and classified the end forces for each component. The fatigue engine evaluated the fatigue life consumption for each component in the model and displayed the results in the results screen. Based on Miner's rule, the stressed component approaches the end of its life when the cumulative damage index exceeds unity. Thus, the end of fatigue life was detected only for the Mast-to-arm connection colored in red with a damage index of 1.116. The stress variation with wind speed for the connection along with the wind speed cycles were plotted and shown in **Figure 12**. The stress life method assumed that the stresses greater than CAFT (18 MPa or 2.6 ksi for this connection) contribute to the fatigue damage accumulation. From the plot,

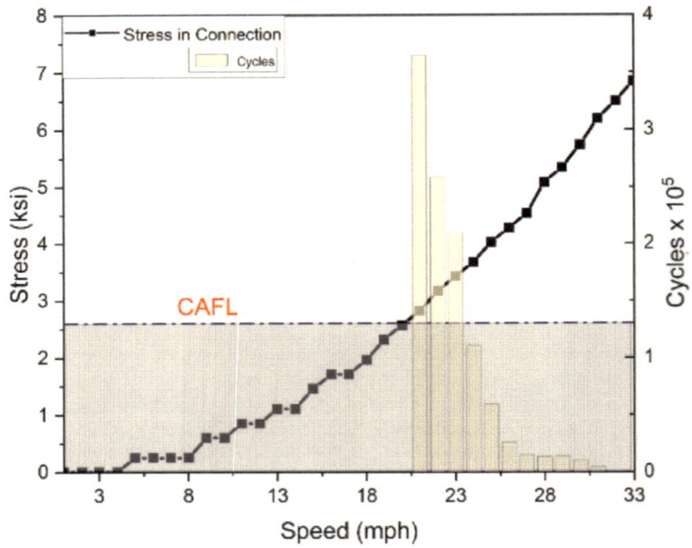

Figure 12.
Stress variation with wind speed in the connection.

only the stresses resulting from the wind speeds (21–33 mph, 34–53 km/h) will cause fatigue damage. The number of cycles to failure associated with each stress is calculated using the S-N equation with A' = 3.29×10^8 ksi. **Table 5** shows a sample fatigue damage calculation as per Miner's rule. The possibility of this complete fatigue crack is likely due to the poor quality of the weld, some defects are very likely to exist resulting in local stress concentration yielding rapid fatigue damage accumulation. Moreover,

Speed (mph)	Stress (ksi)	N_i (Cycle)	n_i (Cycle)	D_i
21	2.82	17421692.22	3,647,732	0.2094
22	3.17	12235811.27	2,582,522	0.2111
23	3.42	9730332.478	2,102,996	0.2161
24	3.67	7864766.583	1,105,565	0.1406
25	4.03	5975640.433	590647.5	0.0988
26	4.28	4981930.229	260,106	0.0522
27	4.53	4196883.267	149,004	0.0355
28	5.08	2969547.639	133981.5	0.0451
29	5.33	2568838.623	138694.5	0.0540
30	5.74	2065094.187	102,642	0.0497
31	6.19	1644141.092	40,464	0.0246
32	6.49	1425231.089	4365	0.0031
33	6.84	1216291.56	405	0.0003
		Sum = 1.116		

Table 5.
Stress and damage in connection associated to each wind speed.

the harsh corrosion environment introduced discontinuities along the weld length which resulted in significantly lower fatigue resistance. It is important to note here that the ancillary structure under consideration was designed 32 years prior to the inspection year when no fatigue design provisions were available. Another important reason to note here is the fact that the geographical area plays a vital rule in the analysis results since the variation in fatigue life is extremely correlated to the difference in wind environment in various sites, Sedgwick County is known for the strong wind records and many ancillary structures have shown a different level of wind-related distress as per KDOT. In addition, the wind-induced fatigue damage was evaluated for all different critical spots in the structure, namely, pole-to-base plate weld connection, chord-to-transverse plate weld connection, anchor bolts, and all the truss members. The fatigue lives of all the previous details were found to be infinite for this particular structure. This is attributed to the fact that at lower wind speeds, all the stresses experienced by these details are below the threshold stress thus no fatigue damage was developed. At the same time, under higher wind speeds, the stresses are higher than the threshold causing a tendency to develop fatigue damage, but the number of cycles is low to an extent not causing this damage to be significant resulting in extended fatigue life.

7. Process generalization

The synthetic wind time histories generated can be extended to any location with updated wind speed records using the same spatial interpolation technique. In cases where the wind speed records are insufficient, temporal interpolation can also be utilized to provide a more comprehensive understanding of wind fluctuations throughout the life of the structure.

While bridge failures can occur due to various reasons, the study of vibrations in a bridge's structural integrity provides insight into how it will perform under severe stress from vibration loading. Similar to highway sign structures, bridge structures are also susceptible to sustained wind loading events that can lead to fatigue failure over time.

To improve the safety and efficiency of bridge structures, effective inspection tools and wind-speed interpolation techniques compliant with AASHTO specifications for fatigue can be developed. This process of calculating fatigue damage in highway sign structures can be generalized to bridge structures. By implementing these techniques, the public's safety and economic stability can be ensured.

8. Conclusion and recommendation

The aim of this chapter is to establish an analytical framework for evaluating fatigue damage in highway sign structures and to develop computer software that can serve as a cost-effective inspection tool. Analytical models were created for a cantilever structure by utilizing past wind event histories to simulate damage in different critical truss components. The key finding from this study is that the pole-to-mast arm connection at the weld is the most critical fatigue detail due to its lower CAFT, which makes it susceptible to fatigue damage accumulation. The prevalence of fatigue failure in flexible highway structures underscores the importance of timely identification of faulty connections. As such, the developed software is expected to have a significant impact on state highway decision-making and inspection inventories.

Acknowledgments

This work was supported by Kansas Department of Transportation, Bureau of structural and geotechnical services through the grant KTRAN: KSU-20-3 (KDOT).

Nomenclature

ξ	Damping ratio
ω_n	Natural frequency of the structure
ω	Excitation frequency
R	Frequency ratio
U_z	the mean wind velocity at height z
f	Specified frequency
U_*	Shear velocity
σ_u^2	variance of the turbulent wind component
γ	Scale factor
t	Time in seconds
K_z	Height and exposure factor
G	Gust factor
I_r	Importance factor
C_d	Drag coefficient
V	Wind velocity
N_i	the number of cycles to failure at i-th stress range
A'	Stress constant associated with structural component
$\Delta\sigma_i$	Member stress
D_i	Incremental damage in member
D	Total damage in member
n_i	Wind speed cycles
σ_{x1}	The axial stress in first quadrant in the main members
σ_{x2}	The axial stress in second quadrant in the main members
σ_{x3}	The axial stress in third quadrant in the main members
σ_{x4}	The axial stress in fourth quadrant in the main members
My	The moment about y-axis
Mz	The moment about z-axis
$2C_3$	The moment arm
r_o	Outer radius
r_i	Inner radius
σ_x	The axial stress in the pole element
Fx	Axial force in the pole
A	Cross sectional area of the pole
I	Second moment of area
t_m	Thickness of the main hollow member
t_{w3}	Thickness of the main members weld
d_b	Diameter of the bolts in the mast-arm connection

G	Half of the width of the connection plate
J	Half of the hight of the connection plate
T	Tension force in the main members
I_{yy}	the moment of inertia along the vertical axis passing through the centroid of the secondary element.
I_{yz}	The product of inertia
I_{zz}	the moment of inertia along the axis perpendicular to both horizontal and vertical axis through centroid of the secondary element
y	Vertical distance from the centroid to the point where the stress is calculated
z	Horizontal distance from the centroid to the point where the stress is calculated
t_p	Thickness of the plate
t_{w1}	Inner weld thickness in the base plate
t_{w2}	outer weld thickness in the base plate
σ_w	Total axial stress in the weld
S_x	Membrane stress in the plate
t_w	Weld thickness in the ring stiffened connection
w_i	Weight function
$\mathbf{J}$	Jacobian matrix
$\mathbf{A}^+$	Moore-Penrose matrix
$S_K(f)$	Kaimal spectrum
f	specified frequency
$\varnothing_i$	Phase angle
k	Surface roughness class

Author details

Khalid W. Al Shboul[1], Husam A. Alshareef[2]* and Hayder A. Rasheed[3]

1 Structural Engineer, Barr Engineering Co., Minneapolis, MN, USA

2 School of Engineering, Colorado State University Pueblo, Pueblo, CO, USA

3 Department of Civil Engineering, Kansas State University, Manhattan, KS, USA

*Address all correspondence to: husam.alshareef@csupueblo.edu

IntechOpen

© 2024 The Author(s). Licensee IntechOpen. This chapter is distributed under the terms of the Creative Commons Attribution License (http://creativecommons.org/licenses/by/3.0), which permits unrestricted use, distribution, and reproduction in any medium, provided the original work is properly cited.

References

[1] AlShboul KW, Rasheed HA, Alshareef HA. Intelligent approach for accurately predicting fatigue damage in overhead highway sign structures. Structure. 2021;**34**:3453-3463. DOI: 10.1016/j.istruc.2021.09.090

[2] DeSantis PV, Haig PE. Unanticipated loading causes highway sign failure. In: Proceedings of ANSYS Convention. 1996

[3] Choi H, Najm H. Fatigue reliability assessment of potential crack initiation of tube-to-transverse plate connections for cantilever sign support structures. Journal of Performance of Constructed Facilities. 2018;**32**:4018002

[4] Al Shboul KW. Deterministic Fatigue Life Simulation of Flexible Sign Structures. Kansas State University; 2022

[5] Hosch IE, Fouad FH. Design fatigue load of sign support structures due to truck-induced wind gust. Transportation Research Record. 2010;**2172**:30-37. DOI: 10.3141/2172-04

[6] Fouad FH, Hosch IE. Design of Overhead VMS Structures for Fatigue Loads. University Transportation Center for Alabama; 2011

[7] Kacin J, Rizzo P, Tajari M. Fatigue analysis of overhead sign support structures. Engineering Structures. 2010;**32**:1659-1670. DOI: 10.1016/j.engstruct.2010.02.014

[8] AL Shboul KW, Rasheed HA, AlKhiary A. Spatial wind speed interpolation using Isoparametric shape functions for structural loading. Structure. 2023;**50**:444-463. DOI: 10.1016/j.istruc.2023.02.037

[9] Al Shboul KW, Rasheed HA. Probabilistic Wind Speed Model for Local Data in all Kansas Counties. Manhattan: Kansas State University Transportation Center; 2022

[10] (ESRI) ESRI. ArcGIS Release 10.1 2021

[11] Ben-Israel A, Greville TNE. Generalized Inverses: Theory and Applications. Vol. 15. Springer Science & Business Media; 2003

[12] Cochran L. Wind issues in the design of buildings. 2012. DOI: 10.1061/9780784412251.ch01

[13] Wang L, Shi W, Li X, Zhang Q, Zhou Y. An adaptive-passive retuning device for a pendulum tuned mass damper considering mass uncertainty and optimum frequency. Struct Control Heal Monit. 2019;**26**:e2377

[14] Wang L, Nagarajaiah S, Shi W, Zhou Y. Study on adaptive-passive eddy current pendulum tuned mass damper for wind-induced vibration control. The Structural Design and Tall Special Buildings. 2020;**29**:e1793

[15] Kaimal JC, Wyngaard JC, Izumi Y, Coté OR. Spectral characteristics of surface-layer turbulence. Quarterly Journal of the Royal Meteorological Society. 1972;**98**:563-589. DOI: 10.1002/qj.49709841707

[16] Iannuzzi A, Spinelli P. Artificial wind generation and structural response. Journal of Structural Engineering. 1987;**113**:2382-2398

[17] ASTM E1049. Standard practices for cycle counting in fatigue analysis. ASTM Standandard. 2017;**85**:1-10. DOI: 10.1520/E1049-85R17.2

[18] Matsuishi M, Endo T. Fatigue of metals subjected to varying stress. Japan Society of Mechanical Engineering Fukuoka, Japan. 1968;**68**:37-40

[19] Bentley. Staad Pro V8i SS6 2016

[20] Alshareef HA, Al Shboul KW, Rasheed HA, Abouelleil A. Analytical-based application software for estimating remaining fatigue life of non-cantilevered sign structures. Engineering Structures. 2022;**262**:114315. DOI: 10.1016/j.engstruct.2022.114315

[21] Ginal S. Fatigue Performance of Full-Span Sign Support stRuctures Considering Truck-Induced Gust and Natural Wind Pressures. MS Thesis. 2003;393

[22] AASHTO. 2015 Interim Revisions to Standard Specifications for Structural Supports for Highway Signs, Luminaires, and Traffic Signals Sixth Edition 2013. 2015

[23] Al Shboul KW, Rasheed HA. Evaluating fatigue in steel cantilevered sign structures under service life wind events through a comprehensive tool for inspection. Journal of Performance of Constructed Facilities. 2023;**37**:4022084

[24] Roy S, Park YC, Sause R, Fisher JW, Kaufmann EJ. Cost-effective connection details for highway sign, luminaire, and traffic signal structures. 2011

Section 2

Bridge Monitoring and Maintenance

Chapter 4

Planning and Repairing Bridges with Robotic and Automation Technology

Vladimir Križaić

Abstract

Today, the maintenance of valuable buildings is carried out continuously and with the help of automated technologies that aim to streamline project activities. The technological innovations are the new machine-robots of Conject technology, and the organisational innovations are the new iterative formula equations of procedures. Thus, the method of dynamic structural programming (DSP) with vectorial methods of norms (VN) and a modified Gaussian S-curve (MGSC) ensures investment planning with modern scientific procedures. In other words, the above automation processes are a development of the rationality of society and provide opportunities for further scientific research in the field of AI systems depending on the wave frequency of their construction.

Keywords: conject robot, vectoral norm, DSP method, DSP COD, MGSC

1. Introduction

Planning is an organisational and managerial function that is essential for the realisation or maintenance of a given project. It enables the optimal and cyclical planning of project activities and the allocation of all resources with the possibility of controlling and regulating the system of project implementation. In practise, this function is often vague and imprecise due to outdated standards. For example, the maintenance of bridges [1] or valuable buildings, for example, is still continuously carried out with new automated robotics, internet communication technology (ICT) and integrated information technology (IIT). That is, digitalised communication with modern management is supported by the IIT management system, but the construction industry systems do not show significant improvements in the economic principles for certain projects. For the sustainability of the transport infrastructure system, it is necessary to respond to the automation of technology with the automation of logistics, that is, with the management of contractor systems in the whole enterprise. The softwareisation of existing organisational and management processes has a problem with static and outdated normative data for the execution of certain activities in the construction process, and the new ICT and IIT technologies do not have a greater impact on the rationalisation of large infrastructure projects. Therefore, it is necessary to replace the old organisation with innovations in organisational management that are translated into new formulas

IntechOpen

and procedural equations, as technology has done with vector systems for design, dimensioning and robotic execution with cyclic iterative processes. Vector and iterative methods are thus proposed to round off the investment using scientific procedures that should help improve economic indicators not only for the company but also for the state institutions. Looking at the problem of maintenance or reconstruction of buildings from a layman's point of view, the statement that it is cheaper to build a new building than to repair an old one is true. It would be the same for bridges, but bridges are valuable objects of cultural heritage and represent the tourist potential of any place. Therefore, the continuous replacement of quickly worn-out structural elements prevents the deterioration of the main structural elements, reducing the essential costs and significantly extending the useful life and profitability of the investment. Every realisation is a production. It is a function of technology, organisation, management and ICT. The technology uses differential equations, while the other components run behind the mathematical modelling of these processes. Therefore, software models and simulations such as BIM and Digital Twins appear today as tools that seek to supplement the lack of decision-making in processes. Thus, the need for vector [2] or parametric [3] modelling and the transformation of POG and PMD documentation into scientific documentation emerges. However, today's ICT digital technology also provides automation of the work, for which the GPS technology was developed, which converts manual collection into automatic data collection in the field and automatically sends it to central data processing in the company's IIS system. However, the key to solving the problem lies not only in automating existing administration but also in standardising new models, construction standards and cost items. This is how the method dynamic structural programming (DSP) or iterative double COD, which creates the iterative differential of cyclical production processes, came into being. Here, mathematical and cybernetic methods are used to transform organisational structures into differential organisational structures as the main components of the AI system. The purpose of automation is softwareisation, that is, the model standardisation of production processes using mathematical and statistical methods to replace routine and dangerous operations with robots. The goal is to establish new vector norms with statistical modelling of modern project planning and control and process automation for all routines, especially dangerous work, to free people from hard and dangerous work. This leads to model standardisation and normalisation of construction activities and at the same time of all economic activities.

2. Conject robot technology of hydrodemoling

Bridge rehabilitation works in the Republic of Croatia are carried out using today's most advanced technology. Modern hydrodemolition technology with Conjet robots and high-pressure hydro pumps in **Figure 1** accelerates the investment cycle in the sustainable development of construction and socio-economic livelihood cycles.

Svedise Innovation Technology's Conjet [4] concrete hydrodemolition robots use a high-pressure water jet of up to 1500 bar (22,000 psi) that travels at a constant speed over the concrete surface, taking advantage of the concrete's permeability to create positive pressure that breaks it up. The automation of the hydrodemolition robots allows operators to easily perform both selective and non-selective removal. Automated Concrete Removal (ACR™) is a concrete removal method that uses robotic technology to remove concrete from structures such as bridges, parking decks, dams and tunnels using high pressure water (hydrodemolition). The hydrodemolition technology ensures that no

Figure 1.
Hydrodemolition of bridge deck with Conjet robots.

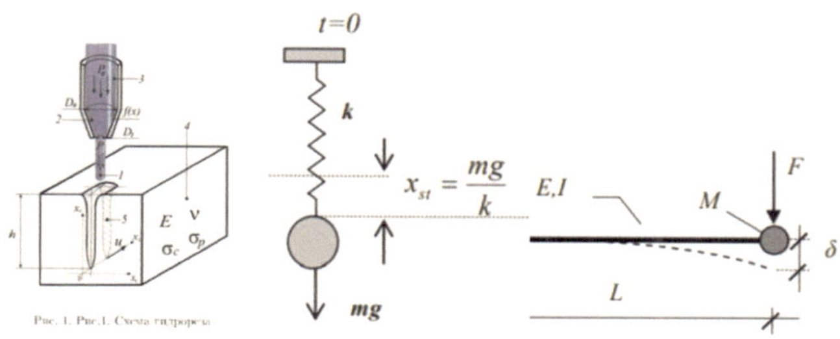

Figure 2.
Schematic representation of the effect of pressures and frequencies [5, 6].

micro-cracks occur during concrete removal and provides an ideal surface for the bonding of new concrete, while the robot ensures the quality and consistency of the concrete removal. The Conjet ACR™ robot removes the same amount of concrete as 20–25 workers with jackhammers. Even better, the ACR™ robot significantly reduces the noise of concrete demolition. Djelovanje mlaza na konstruktivne elemnte stavara prostorne sile koje se trigonometrijom proraunavaju u vektorskom prostornom i frekvencijskom polju. The Conjet robot with a pump over 1000 bar or > 100 N/mm^2 and a remote-controlled programme automatically hydrodemolishes concrete, which is usually designed for strength C 30/37 or 30 N/mm^2. The generation of pressure by the pump three and more times than necessary to destroy the concrete is demonstrated in **Figure 2**.

3. Vectoral norms or effects

The methods of technical standardisation [7] and measurement of time consumption are the methods of timekeeping, photo-inspection, recording, ongoing observations and technical records. Using the method of the work diary, that is, technical records, the statistical distribution curve of the effect of the robot in **Figure 3** was

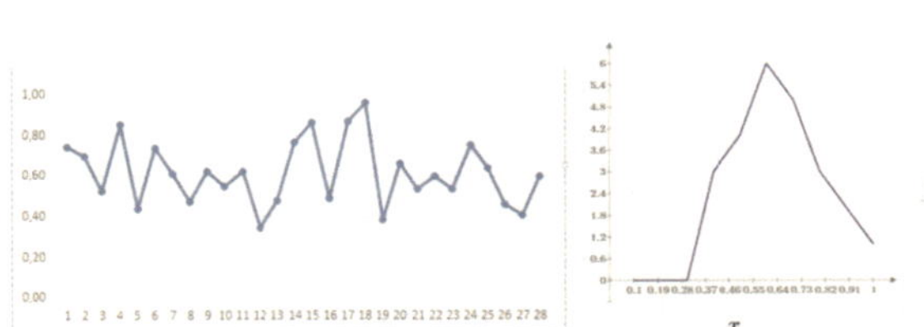

Figure 3.
Discrete frequency of the effect partial and cumulative by effect from monitoring the variables of Conject robots 364 operations at level 5/7.

Date	RN	pump	robot	machinist 1	machinist 2	operation	shift	depth	width	length	m+	effective work hours	norm hours
28.11.2007	111.752	364	364	Kovačić.Z	Ivšić	staza	2	0.2	0.3	35.7	2.14	11	0.19
17.11.2007	111.752	364	364	Šobak	Osti	popravci	1	0.06	2	10	1.20	11	0.11
27.11.2007	111.752	364	364	Glavica	Butorac	popravci	1	0.06	2	26	3.12	7	0.45
12.11.2007	111.752	364	364	Damjanović	Drljan	ploča 5/7	2	0.06	2	24	2.88	3	0.96
13.11.2007	111.752	364	364	Palijan	Butorac	ploča 5/7	1	0.06	2	55	6.60	11	0.60
13.11.2007	111.752	364	364	Damjanović	Drljan	ploča 5/7	2	0.06	2	53	6.36	11	0.58
14.11.2007	111.752	364	364	Palijan	Butorac	ploča 5/7	1	0.06	2	55	6.60	11	0.60
14.11.2007	111.752	364	364	Miličević	Krivokuča	ploča 5/7	2	0.06	2	51	6.12	11	0.56
15.11.2007	111.752	364	364	Palijan	Butorac	ploča 5/7	1	0.06	2	66	7.92	11	0.72
15.11.2007	111.752	364	364	Miličević	Krivokuča	ploča 5/7	2	0.06	2	60	7.20	12	0.60
16.11.2007	111.752	364	364	Palijan	Butorac	ploča 5/7	1	0.06	2	55	6.60	11	0.60
16.11.2007	111.752	364	364	Miličević	Krivokuča	ploča 5/7	2	0.06	2	34	4.08	10	0.41
17.11.2007	111.752	364	364	Miličević	Krivokuča	ploča 5/7	2	0.06	2	50	6.00	10	0.60
18.11.2007	111.752	364	364	Butorac	Osti	ploča 5/7	1	0.06	2	48	5.76	11	0.52

Table 1.
Schedule of work of machinists on robots and pumps for water demolition.

sum	24	1	0,646
t	*n*		***E*(*x*) = 0,65**
x	*y*	Px	*x**Px
0,1	0	0,00	0,000
0,2	0	0,00	0,000
0,3	0	0,00	0,000
0,4	3	0,13	0,050
0,5	4	0,17	0,083
0,6	6	0,25	0,150
0,7	5	0,21	0,146
0,8	3	0,13	0,100
0,9	2	0,08	0,075
1	1	0,04	0,042

Table 2.
Effect frequencies of E(x) *robot 364 on the Zagreb-Zaprešić bridge.*

determined from the analytical data **Table 1** by observing and recording the process of hydrodemolition at the bridge in Zagreb over the Sava.

These measurements were transformed into a discrete function, which is approximately replaced by the exponential Weibull distribution derived from the Erlang distribution [8], which is typical for work and service activities. The given discrete plotted example was transformed into a function of the event frequency density of the robot work effect so that the effect and the norm, that is, the expected duration of the process, are determined.

In column **Table 2**, $P(x)$, that is, the probability of occurrence of the robot's effect frequency, was defined to determine the mathematical expectation of the statistical effect variable for a given robot 364 activity. It follows that $E(x)$ (Eq. 1) is 0.646, that is, 0.65 m^3/h for the robot and the pump operated by the **Table 1** mechanism.

$$E(x) = \sum_{X} x * P(x) \tag{1}$$

The second method of functional connection of the effect using the modified Erlang curve gives an approximate result (Eq. 2).

$$y(x) = \lambda(\lambda x)^k e^{-\lambda x}/k! \tag{2}$$

$E(x) = \int_{-\infty}^{\infty} x(y_x)dx$ The mathematical expectation value of the Erlang distribution is known by k and λ (Eqs. 3 and 4).

$$E(y_x) = \int_{-\infty}^{\infty} x(y_x)dx \tag{3}$$

$$E(x) = (k+1)/\lambda \tag{4}$$

The parameters k and λ are discrete values of a sample for whose values the Erlang function approximates the distribution of observations of the effects of a given robot. By using the Matcad tool and assigning certain parameter values, we define and observe whether the curve corresponds to our selective distribution. The modified Erlang curve has an exponent for the variable x $k-1$ instead of k (Eq. 5). The variable x is in the interval $[1,2,-10]$, $k = 6$ and $\lambda = 0.96$, and the Erlang curve in **Figure 4** is defined as $E(x) = 0.7\ \text{m}^3/\text{h}$, that is, the expected effect of the observed hydrodemolition robot.

$$f(x) = 1/k!\left(\lambda^{k+1}x^{k-1}e^{-kx}\right) \tag{5}$$

Modified Erlang work allocation of machinists on hydrodemolition robots and pumps.

It can be seen that the mathematical expectation of the standard for the definition of a continuous event is higher than for the definition of a discrete event, since the continuous line fills the gaps of the discrete distribution (Eq. 6).

$$E(y_x) = \int_{-\infty}^{\infty} x(y_x)dx > \sum_i \left(x_i y_i / \sum_i y_i\right) \tag{6}$$

Thus, the approximate solution for the expected performance of the robot can also be determined by the third common data correlation method, the Gaussian sum of least squares method [9, 10]. The given curve is parabolic and resembles the Gaussian distribution density function (Eqs. 7–10), with input variables and matrix b and matrix kooeficijent for robot performance. The matrix variables for the given Gaussian system are the matrix t with the frequency of the effect and the matrix x with the

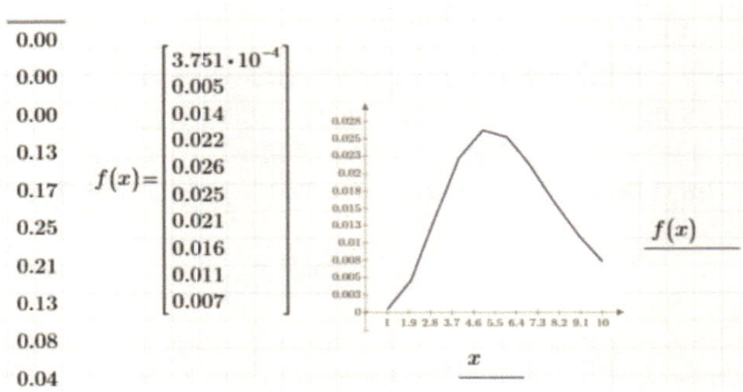

Figure 4.
Modified Erlang labour distribution of machinists on hydrodemolition robots and pumps.

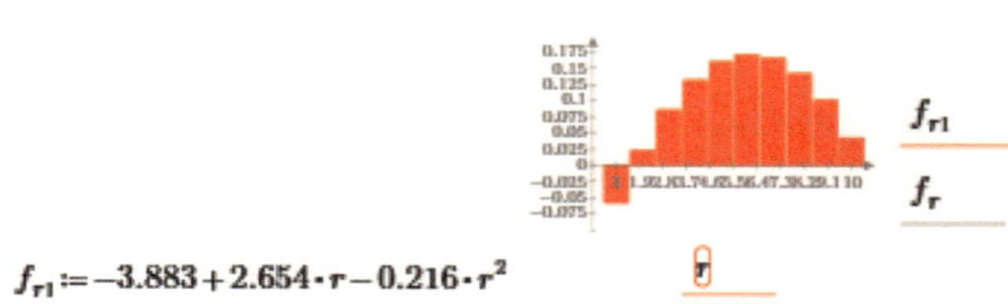

Figure 5.
Modified parabolic distribution of machinists' work on hydrodemolition robots and pumps.

numerator of the effect value of the machine, which is replaced by a vector with unit numbers of the effect multiplied by 10. The number of measurements is n = 10 and the result is **Figure 5**.

$$x = \begin{bmatrix} 1 \\ 2 \\ 3 \\ 4 \\ 5 \\ 6 \\ 7 \\ 8 \\ 9 \\ 10 \end{bmatrix}, t = \begin{bmatrix} 0 \\ 0 \\ 0 \\ 3 \\ 6 \\ 4 \\ 5 \\ 3 \\ 2 \\ 1 \end{bmatrix} \tag{7}$$

$$b = \begin{bmatrix} b1 \\ b2 \\ b3 \end{bmatrix} \; b1 = \sum_{x=1}^{n} x, b2 = \sum_{x=1}^{n} xtb3 = \sum_{x=1}^{n} x^2 t \tag{8}$$

$$\begin{bmatrix} n & \sum_{x=1}^{n} x & \sum_{x=1}^{n} x^2 \\ \sum_{x=1}^{n} x & \sum_{x=1}^{n} x^2 & \sum_{x=1}^{n} x^3 \\ \sum_{x=1}^{n} x^2 & \sum_{x=1}^{n} x^3 & \sum_{x=1}^{n} x^4 \end{bmatrix} \times \begin{bmatrix} X_1 \\ X_2 \\ X_3 \end{bmatrix} = \begin{bmatrix} b1 \\ b2 \\ b3 \end{bmatrix} \tag{9}$$

$$X = bA^{-1} \qquad X = \begin{bmatrix} -3{,}883 \\ 2{,}645 \\ -0{,}216 \end{bmatrix} \tag{10}$$

The variable r replaces x, and by correcting the obtained equation, we approximate the quadratic equation with the product 1/25 and get the distribution in **Table 3**. From this approximation, it follows that $E(x)$ is 0.619, or 0.6 m^3/h, taken for the Conject 364 robot.

sum	**18,1**	**1**	0,619
t	*n*		***E*(*x*) = 0,6**
x	*y*	Px	*x**Px
0,1	0	0,00	0,000
0,2	0,5	0,03	0,006
0,3	1.5	0,08	0,025
0,4	2,3	0,13	0,051
0,5	2,8	0,15	0,077
0,6	2,9	0,16	0,096

sum	**18,1**	**1**	0,619
0,7	2,8	0,15	0,108
0,8	2,3	0,13	0,102
0,9	2	0,11	0,099
1	1	0,06	0,055

Table 3.
Frequencies of the effect with E(x) *of robot 364 on the Zagreb-Zaprešić bridge when returning from a quadratic function.*

Sort the given database with various queries from the given variables that influence the effect is most often the depth of hydrodestruction, which depends on the brand of concrete. Before the hydrodemolition, the bridge was divided into fields, and the strength of the fields was determined using a sclerometre (**Table 4**).

Discard the min and max performance data and brand of concrete, the proportionality of the dependence of these data is visible in **Figure 6**.

His is the definition of the equation used to determine the dimensional coefficient kd, the concrete strength kdc. The base or unit is taken for C 30/35. The solution of the

X,C \ T	T1	T2	T3	T4	T5	T6
$\underline{C}$	34,2	40,32	34,2	38,88	31,32	34,2
$\underline{X}$	34	37,4	33,4	36,6	32,4	34
efect	0,96	0,59	0,58	0,66	0,5	0,6
28	2	0	2	0	1	0
30	1	0	1	0	4	2
32	1	0	2	2	2	3
34	1	1	2	0	1	1
36	1	5	1	3	0	3
38	1	0	0	3	1	2
40	2	4	2	2	1	0

Table 4.
Frequencies of the strength of the concrete on the Zagreb-Zaprešić bridge by field.

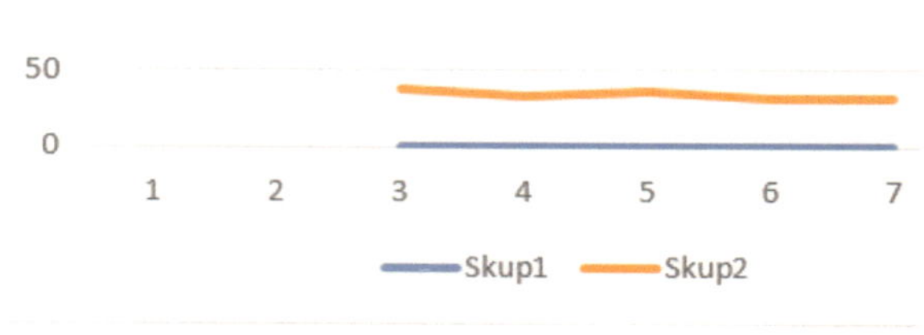

Figure 6.
Hydrodemolition of bridge deck with Conjet robots.

$f_{r1} := 68.25 - 52.3 \cdot r$

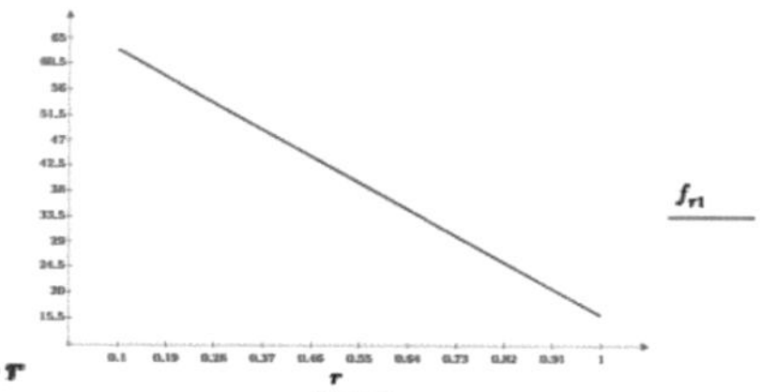

Figure 7.
Dependence of the thickness of the bridge deck on the panel thickness and the robot performance.

curve as a function of the action variable t and the concrete strength variable x is obtained by linear Gaussian regression (Eqs. 11–13) Input variables with matrix b and linear matrix systems are defined in **Figure 7**.

$$x = \begin{bmatrix} 0,59 \\ 0,69 \end{bmatrix}, t = \begin{bmatrix} 37,4 \\ 32,17 \end{bmatrix}, b = \begin{bmatrix} b1 \\ b2 \end{bmatrix}, b1 = \sum_{x=1}^{n} x, b2 = \sum_{x=1}^{n} xt \tag{11}$$

$$\begin{bmatrix} n & \sum_{x=1}^{n} x \\ \sum_{x=1}^{n} x & \sum_{x=1}^{n} x^2 \end{bmatrix} \times \begin{bmatrix} X_1 \\ X_2 \end{bmatrix} = \begin{bmatrix} b1 \\ b2 \end{bmatrix} \tag{12}$$

$$X = bA^{-1}\ X = \begin{bmatrix} -3{,}883 \\ 2{,}645 \\ -0{,}216 \end{bmatrix} \tag{13}$$

This results in the functional dependence of the dimensional coefficient, that is, the concrete strength kdc, represented by the assumed linear function (Eq. 14). With the given coefficients, kdc is decimally reduced by the numerator n, which follows the series of concrete strengths of + C 5.

$$k_{dc} = -0,1n + 1,2 \tag{14}$$

Thus, the numerator n has the interval [1–5], bat C [30, 35–50]. Through the one-year study of robotic work on over 1700 m^3 of concrete and over 3200 hours effectively spent, it was found that the shift work does not have much effect, while the width of the work is equal to the depth. The depth observations show that they depend on the brand of concrete. Thus, it can be seen that for $d > 20$, the effect is approximately constant, and they are beams of stable concrete brand C 40/50, while for smaller thicknesses for slabs where the concrete has failed (the effect of salt on the bond between the reinforcement and the concrete), the dispersion of the effects is approximately equal to that of the strength of the concrete. The mathematical expectation technically lies approximately in the middle of the Erlang distribution, so it is satisfactory to use this point as the average mathematical expectation effect of the robot for calculation and planning procedures. By testing the given Erlang distributions with the chi-square test for $\alpha = 0.05$ and 2 degrees of freedom, $\chi 2 < \chi 2\ \alpha$, the given curves are valid for a random sample, since the deviations of the empirical distribution from the Erlang distribution are not significant in a tabular representation of the mathematical performance expectation of the processed robots in **Table 5**.

Robot	$E(x)$			X	$\chi^2 < \chi^2_\alpha$
		$d \leq 20$ cm	$d > 20$ cm		5,99
361	0,7	0,8	0,48	0,54	5,75
364	0,6	0,65	0,45	0,56	2,84

Table 5.
Mathematical expected performance of all robots on the Zagreb-Zaprešić bridge in 1 month.

With several considerations of the mathematical expectation $E(x)$ through **Table 2** and the simplified linear method of the arithmetic mean, one obtains the average effect of the resource, which becomes a vector effect variable in **Figure 8**.

Substitutes for the graphical representation are the equations of the base and other graphs, that is, the water destruction of plates, bars or vertical structures. Thus, the equation of the basic graph for water destruction of plates (Eq. 15) is.

$$U_{k,r,ko,kd} = k_{dc} k_o U_{k,r} \tag{15}$$

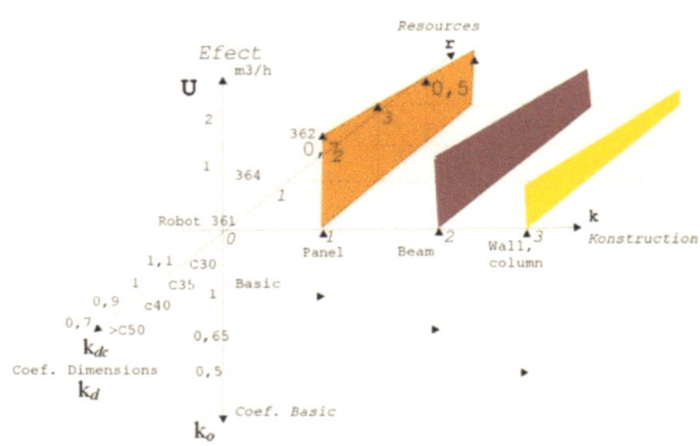

Figure 8.
Vectorial effect of water destruction by Conjet robots.

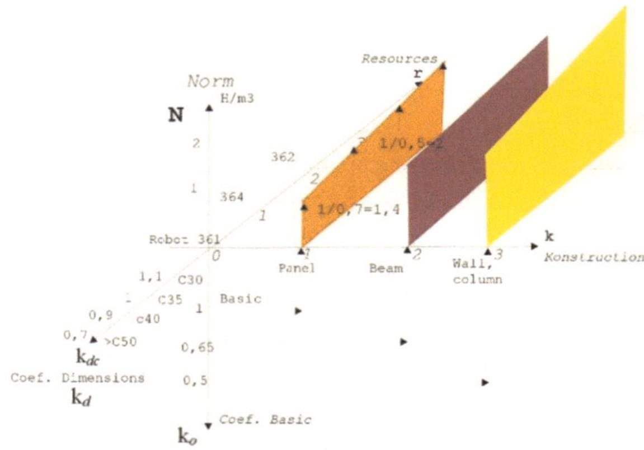

Figure 9.
Vector norm of water destruction by Conjet robots.

where k_{dc} is the coefficient of the construction dimension C for the of concrete, the base coefficient ko which is a function of the structural element, and the equation of the effect U_{kr} which is a function of the resource r and the structural element k. For the base effect, the listed coefficients are uniform so that the equation takes the form of a function of the water destruction effect of the C35 plate. The reciprocal of the effect [m^3/h] gives the standard [h/m^3] (Eq. 16) (**Figure 9**).

$$N = \frac{1}{U_{k,r,ko,kd}} = 1/k_{dc}k_oU_{k,r} \tag{16}$$

It can be seen that the slab is easier to hydrodemolish than the beam and especially the wall due to the effect of the forces acting most strongly in the direction of gravity.

4. Planning with the DSP method

4.1 Software tools for planning

For the realisation of the project, it is important to prepare the work, that is, to plan the development of the production process of the object and the management to regulate the production processes, which are the function of the technology, the organisation, the management and the ITC system with daily system management, which enables the daily regulation of the system (**Figure 10**) [11, 12].

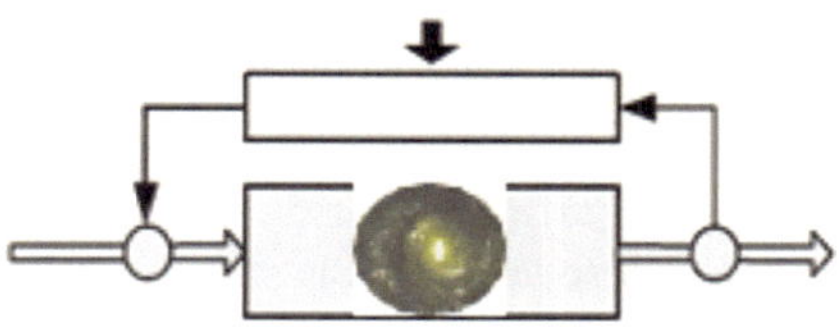

Figure 10.
Regulating of the system.

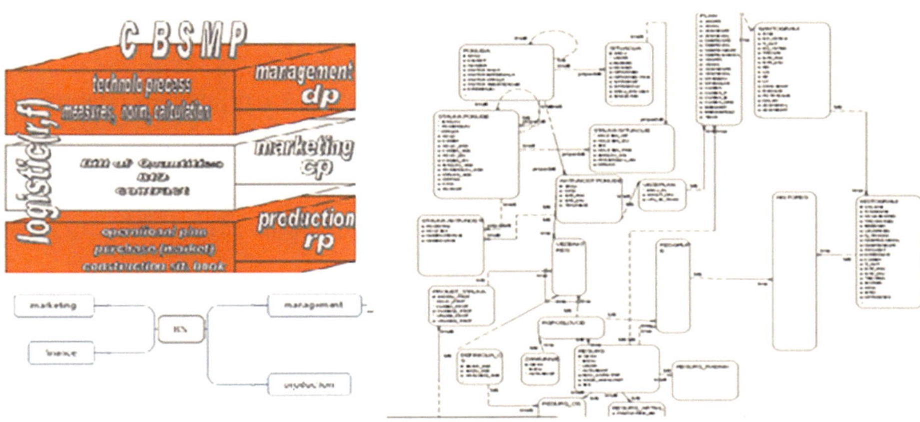

Figure 11.
IIS construction companies with a vector data model structure.

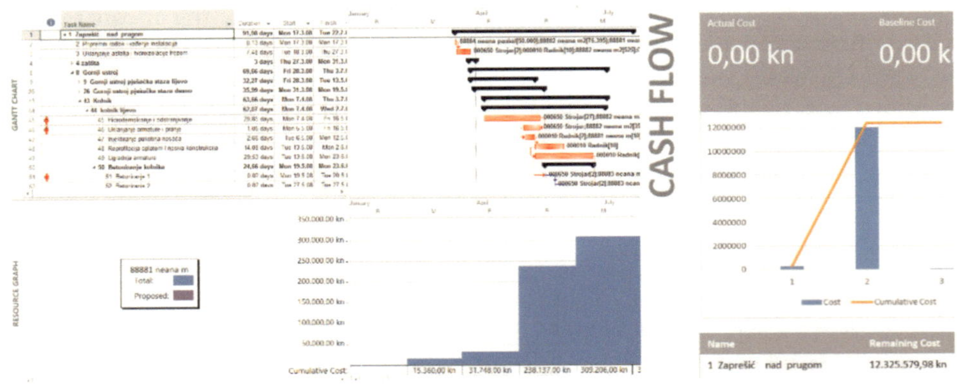

Figure 12.
Project monitoring and cash flow report of the Zagreb-Zaprešić bridge.

Software tools for planning outside the IIS system (**Figure 11**) are developed on the method CPM from Superproject to Perimavera and MSP. Nevertheless, planning in construction is at a low level due to outdated standards. Therefore, the introduction of dynamic or parametric vector standards is recommended to increase the accuracy of planning and thus the costing of projects. Therefore, relational databases create a vector organisational structure [13] of the company, which is introduced into the matrix organisational structure on the third axis, that is, the resource variable stored in the database with histograms.

Integration is done using the foreign software tools for using ODBC databases or the new XML Internet technology. There are also other possible solutions based on operational research through software combinatorics. However, the graphics functions used by indigenous engineers are inferior to those mentioned above. Therefore, Microsoft graphics have been integrated into indigenous software and a report produced in combination with Microsoft tools is shown in **Figure 12**.

The system allows monitoring of resource efficiency by activity and on a daily basis through the presented data model. Monitoring of normative and financial indicators is enabled alternately, which leads to an increase, that is, streamlining, of the control of resources of projects that were previously monitored on a periodic financial basis, from now on on a daily basis and in the future on the basis of normative indicators. This addition is beneficial not only for a company but also for society (the wider community). A rational decision-making process combined with the new scientific approach to the component that is crucial in any society, namely, the value of human labour, would lead to more humanistic outcomes. Once the standards are objectively set, humanistic solutions are yet to be achieved as we enter the realm of humanism and humanity, qualities we lack today. The DSP method paves the way for the standardisation of models through the production of bills of quantities, draft bids and standards.

4.2 Model's standardisation of the bid construction

By harmonising the description of the item in the tender construction, that is, the text of the bill of quantities, that is, the standard, a unique code is created to define the construction product. A product record is formed by layers and distribution of complex processes to procedures, and the construction production

Figure 13.
Standardisation of activities: elements of the description of delivery items in construction [14].

record is modelled by the combinatorics and linking of these records through model standardisation (**Figure 3**). A record of the elements is given by the project MSP (Eqs. 17 and 18), definition of the structure with a given resource and a description of the operational elements, from activities, processes and procedures for construction production and other more advanced production to the movement of robotisation of production, the equation of model standardisation of production (**Figure 13**).

$$MSP = \sum A = \sum Pe = \sum O = \sum Pd = \sum M \tag{17}$$

$$MSP = \sum p = \sum c = \sum r = \sum d \tag{18}$$

Here, MSP forms the left side of the picture with the variables activity A, process Pe, operation O, procedure Pd and movement M, while p is the project, c the construction, r the resources and d the dimension [15]. Modular and variant components can simulate all practical processes and operations up to the procedure through CPS, MindJet graphics technology or the software method DSP.

4.3 DSP COD

DSP code is created by combining dynamic programming with object or structural programming. It is very interesting because resource consumption in multiple processes can be recorded simultaneously, assuming that the given capacity is used. When we consider a multi-process system, we can define a set of vectors x. The optimisation determines which process we need to allocate more resources to in order to obtain a greater profit. To obtain this solution, it is necessary to write down all possible distributions of resources through all processes, that is, to write down and calculate all possible paths. The recordings can be made using cybernetic equations (Eq. (19)).

$$F(X_j) = \sum_{j=1}^{n} g_j(x_j), \quad S = \sum_{j=1}^{n} x_j \quad 0 \leq x_j \geq S \tag{19}$$

That is, by the method of recurrent equations, whose characteristic is the iteration of the functional equations of the state of the system f_n (S_n) in the sum of the observed function g_n and the function of the previous state $f_{n\text{-}1}$ $(S - x_n)$ with all possible changes in the value of the variables in the given functions or processes (Eq. (20)).

$$f_n(S_n) = max_{\,0 \leq x_n \leq S}\big(g_n(x_n) + f_{n-1}(S - x_n)\big) \tag{20}$$

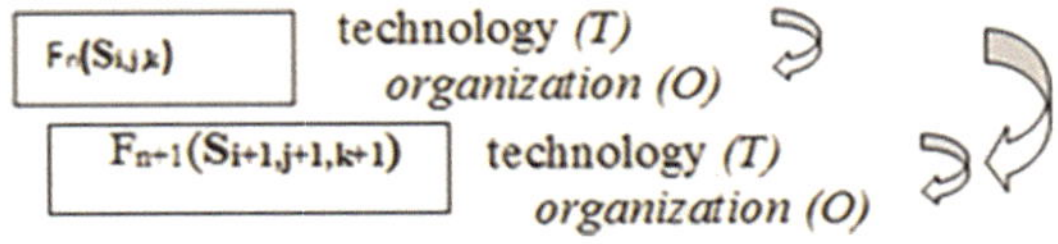

Figure 14.
DSP vector iteration organisational differential.

For multidimensional vector dynamic programming, the equations are identical, only with more variables (x,y). The organisational differential is marked as follows in **Figure 14**.

The mathematical methods of logic and inductive optimisation and the programming methods define the dual DSP structure. Baumholz plus the induction equation, that is, dynamic programming with object programming, defines the idea of the DSP equation to model the production process, that is, construction bid construction with DSP code (5) and TROSKO code. By linking and extending the above models with iteration software technology, a DSP model has been developed to help solve the problem of defining production and products. Thus, the product of the bid construction and the item element are defined as MSP = Σ Fn(S) = f(T,O), as a function of the record of technology (design) and execution organisation, and T ≈ O = f(A,R,D), that is, as a function of the variables *A*—activities, *R*—resources and *D*—dimensions of design and resource performance. The greatest influence is the standardisation of records of bid or bill of quantities items as a recurring form (Eqs. 21 and 22).

$$BC = U\left[f_n(T_n, O_n) = f_{n+1}(T_{n+1}, O_{n+1})\right] \tag{21}$$

or

$$DSP\ COD = U\left[f_n([A|R|D|||]) = f_{n+1}([A\ |R|D|||])\right] \tag{22}$$

While substituting identical structures T O the iteration DSP CODE of the product (**Figure 15**) results.

4.4 Planning with the code DSP

Planning with the DSP code uses a combinatorial Eq. (15) to define all paths in product creation. The DSP code is a universal equation that can be used for all equations, including CPM equations or DSPiP in **Figure 16**.

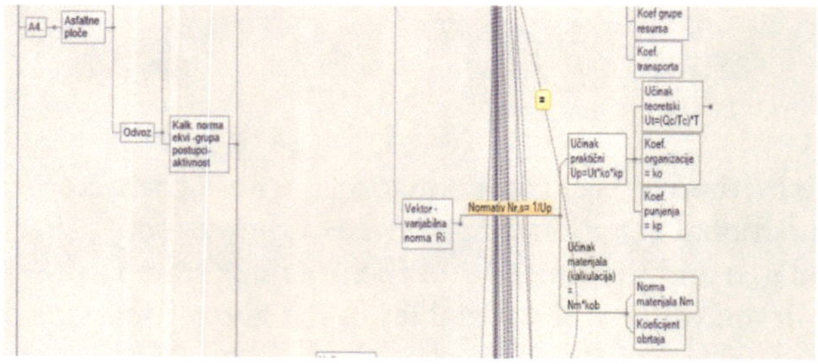

Figure 15.
BC-DSP COD in the MindJet software.

DOI: http://dx.doi.org/10.5772/intechopen.112253

Figure 16.
DSPIP in the Delphy software.

5. Simulations with MGSC for water breakage

More recently, probabilistic forecasts of project performance and the use of stochastic S-curves with a stochastic S-curve generation software package and simulation approach have defined the dispersion variance of finances and time of the modified Gaussian S-curve (MGSC) project as a function of the density distribution of cost and time. Expected monetary value (EMV) or S-charts are widely used today and are complemented by functional MGSK [16, 17] (Eq. 23), (**Figure 17**).

$$skvGr(x,T) = \lambda kv \cdot \int_0^x \frac{1}{(a \cdot T + b) \cdot \sqrt{2 \cdot \pi}} \cdot e^{\frac{-(x-\mu)^2}{kv \cdot (a \cdot T + b)}} dx \qquad (23)$$

The levelling and fitting of the curves, that is, the modification of the Gaussian curve, refers to the introduction of a constant kv parameter in the value of 10,000 units. The investment value for the reconstruction of the Zagreb-Zaprešić bridge is visible in the MSP case stream and amounts to HRK 12325579. In the software budget, the cost axis is in thousands, so $T = 1233 \times 10^4$. The number of working

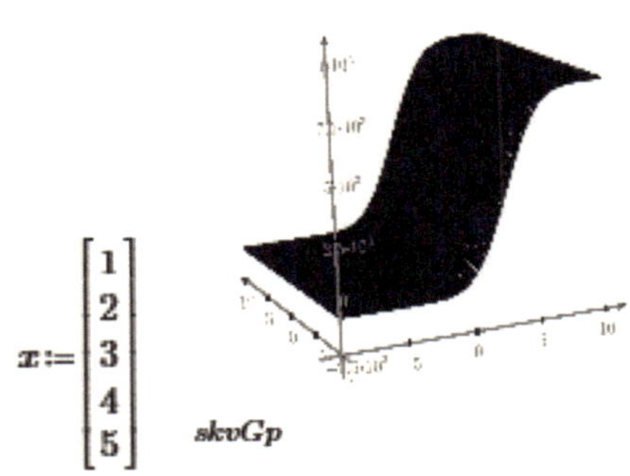

Figure 17.
MGSK project Zagreb-Zaprešić.

$$skvGr(x,T)=\begin{bmatrix}1.223\cdot 10^3\\3.28\cdot 10^3\\4.555\cdot 10^3\\3.28\cdot 10^3\\1.223\cdot 10^3\end{bmatrix}$$

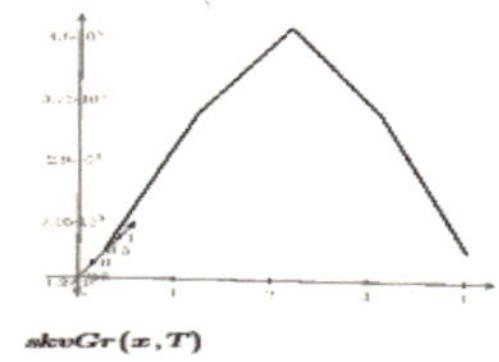

Figure 18.
Simulation of project management with MGSC for the Zagreb-Zaprešić project.

months for the project is $n = 5$, which defines the variable x, and the expectation on the variable x is $x/2$. The constant variable kv is 104, and λkv is assumed to be 3.375.

The Dis S-curve has monthly rates for the investment distributed within 1.2 to 4.5 million in **Figure 18**, which is much more optimal for the S-curve of the Gantt chart than the flow curve of the case from the SME Gantt chart.

6. Automation and AI systems for hydrodemolition

By linking the production engineering structures to the organisational process and creating a functional definition of the structures with a cybernetic record of all combinations, we obtain the DSP -e model of the product definition, that is, the cybernetic system [18–21] of the product definition in **Figure 19**.

By defining the structure of the product with technology T and organisation O, the equation of the product and the cost of the code are defined, which allows the logical function of defining a specific action in the project. In this way, we arrive at the

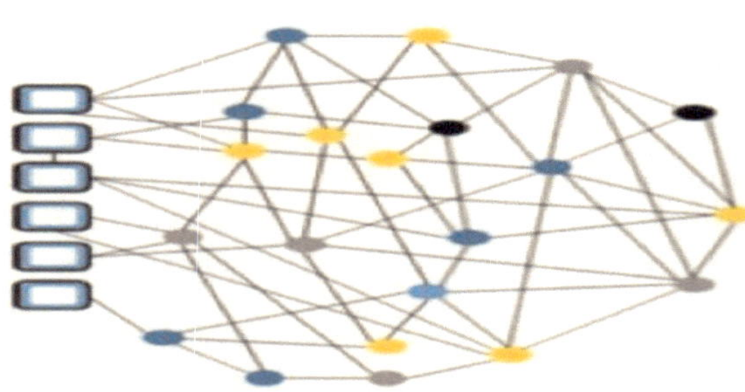

Figure 19.
Sayber product system [18].

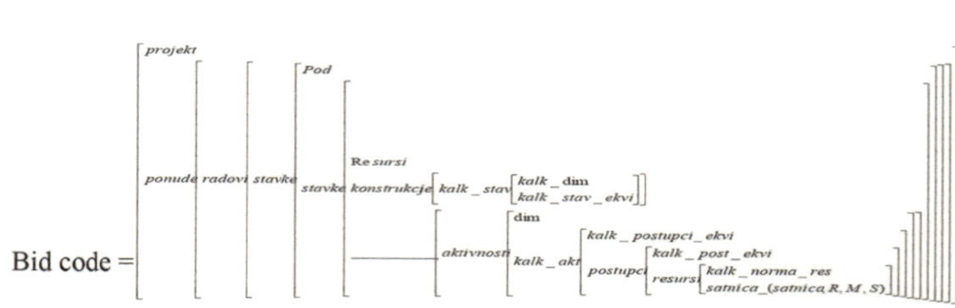

Figure 20.
Sayber product system.

automation or fragmentation of parts of the product defined by the bid code (Eq. 24) (**Figure 20**).

$$Constructionbid = f(T, O, M, ICT) \tag{24}$$

The equation defines the iteration equation of the Sayber system to define the construction bid construction, that is, the product, that is, the execution object. By supplementing the dynamic programming equation with structural or object programming, DSP CODE, or the DSP method of product definition, is developed. The MGSC statistical method, the least squares method and the structural modelling and combination method are used to optimise the creation of a proposal design. The given equation primarily enables the calculation of the quotation construction, and with the filling of the unit price, it enables the system to calculate the critical planning path using combinatorial records of all states and rotational links. Thus, linking the organisational component of the standard or effect to the concrete strength (Eq. 25) opens the field for frequency calculation on the bridge during hydrodemolition, because the frequency of the structure (Eq. 26) is also a function of the concrete strength variable, which is replaced by the strength function as a function of the robot's performance using Eq. (24). Thus, the frequency within the audible radio waves [22, 23] (Eq. 27) (**Figure 21**), is obtained as a function of the thickness of the structure and the strength of the concrete.

$$f_c = -52,3x + 68,25 \tag{25}$$

$$f_x = \frac{\sqrt{\frac{\sigma 3I}{m}}}{2\pi} \tag{26}$$

$$f_x = \frac{\sqrt{\frac{f_c 3I}{m}}}{2\pi} \tag{27}$$

If one enters the values for the standard physical dimensions of the mass of the structure for reinforced concrete 25 kN/m^3, the moment of inertia of the slab 10 cm high and the Ludolf number constant from the circumference of the circle and the strength ab. Construction replaced by the power equation, the result is a matrix of the bridge frequency with about 1000 Hz. However, larger absorption plates with a height of up to 90 cm also generate radio waves.

The given equations connect organisation and technology, that is, economic and physical processes. If you further complete the method DSP by planning a three-level set of permutations and variations, the form of the equations shows the software code of loops over the variables within the matrix structure, which has no standard

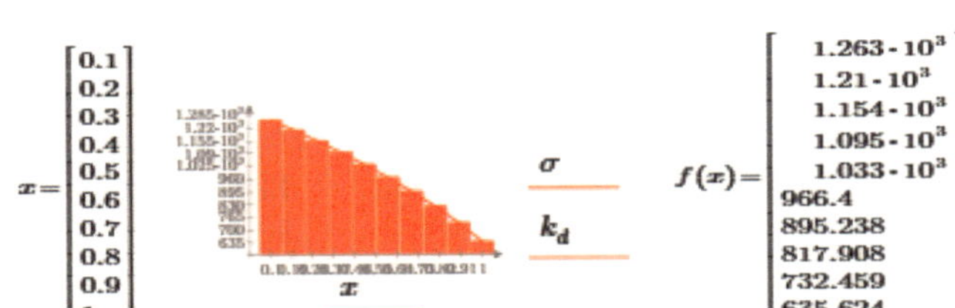

Figure 21.
Frequency of the bridge plate from the hydrodemolition action of the Conject robot.

mathematical properties but is software mathematics. Combinatorics creating a data set of all Croatian lottery combinations opens up the possibility of defining permutations, variations and combinations given by the equation (Eq. 28) in a few minutes. Such combinatorial iterative equations are suitable for computers. Permutations can also be represented graphically by adding the next number to all intermediate columns and the first and last columns of the previous data set. An iterative software equation defines a series of records of all combinations or variations using a matrix representation [24].

$$K = \left[\begin{array}{c|c|c|c} \underset{\overleftarrow{n_{max}}}{n_1 \rightarrow k_{max} - 1} & \underset{\overleftarrow{n_1 - k_i}}{n_2 \rightarrow 1} & \underset{\overleftarrow{n_1 - k_i}}{n_i \rightarrow n_i + 1 \rightarrow \leftarrow n_k - 1} & \underset{\overleftarrow{n_{max}}}{n_k \rightarrow n_1 + 1} \end{array} \middle\| \right] \quad (28)$$

Today, Mathcad has also developed greatly and can also write matrices graphically using software mathematics so that such matrices or determinants can be called software matrices.

7. Conclusions

Optimal investment and maintenance of bridges are the result of new techniques in organisation, management and ITC methods. Operational research with software technology is developing new mathematical models that simulate investment costs for optimal, rational management of capital projects. Any improvement in the organisational or technological cycle is an important component for the profitability of the investment. This requires a new model of standardisation with the vector form of VN-KOD standards defined by DSP code that enables the automation of building products from project documentation to production. Combinatorics and functional linkage connect all branches of science. Just as theory complements practise and vice versa, organisation appears as progress in technology and vice versa. It is a path to AI systems and a great contribution of digitalisation to daily management, that is, the creation of profit within software-mathematical organisational problems. This leads to system automation at the system level of physical processes and controls and the establishment of intelligent project management [25] with the ambition to define AI [26] with simultaneous management. Thus, Android technologies communicatively unite the concept of the fourth industrial revolution through the cloud of the internet and virtual technology [27]. AI systems with model standardisation and vector normalisation of production processes through mathematical and statistical methods approach precision reduce risk [28] and increase sustainability of bridges through channel maintenance theory [29]. DSP Iteration programming with HV normalisation enables chipping, that is, automation and robotisation of large infrastructure investments in a circular sustainable economy in the economic sense with an increase in the principles of frugality, productivity and profitability. In a humanistic sense, AI systems free workers from dangerous and difficult work and open up DSP and cost-effective constant modelling and simulation of model standardisation.

Conflict of interest

The authors declare no conflict of interest.

DOI: http://dx.doi.org/10.5772/intechopen.112253

Notes/thanks/other declarations

Thank you IntechOpen.

Author details

Vladimir Križaić
Politechnic of Medjimurje, Sustainable Development, Sustainable Construction, Čakovec, Croatia

*Address all correspondence to: vladimir.krizaic@gmail.com

IntechOpen

© 2024 The Author(s). Licensee IntechOpen. This chapter is distributed under the terms of the Creative Commons Attribution License (http://creativecommons.org/licenses/by/3.0), which permits unrestricted use, distribution, and reproduction in any medium, provided the original work is properly cited.

References

[1] Križaić V. Bridge maintenance automation, CCC 2016, Budapest, Hungaria. Procedia Engineering. 2016; **164**(2016):143-149. DOI: 10.1016/j.proeng.2016.11.603

[2] Abdelhady MI, Abdelgadir AK, Al-Araimi F, AL-Amri K. Using algorithm in parametric design as an approach to inspire nature in architectural design, Springer Nature. 2022. DOI: 10.1007/978-3-030-97196-0_10

[3] Križaić V. Application of norms models with vectoral system in Construkction projects. Journal of Civil Engineering and Architecture, USA. 2014;**8**:722-728 ISBN 1934-7359

[4] Basic terms. The Widest Range of HydrodemolitionEquipment [Online] 2023. Available from: http://www.Conjet.com

[5] Raduka V. Mehanika II-dinamika, FGZ, Zagreb. 2003

[6] Ivanšić I. Numerička matematika, Element, Zagreb. 1998

[7] Marušić J. Organizacija građenja, Zagreb. 1994

[8] Ivković Z. Matematička statistika, Naučna knjiga, Beograd. 1980

[9] Vukadinović S. Elementi teorije verovatnoće i matematičke statistike, NIGRO - Privredni pregled, Beograd. 1978

[10] Newbold P, Carlson W, Thorne B. Statistika za poslovanje i ekonomiju, Copyryght – MATE, Zagreb. 2010

[11] Križaić V, Hranj D. Daily monitoring by IIS, 11th OTMC, Dubrovnik. 2013. ISBN 978-953-7686-03-1, str. 26

[12] Križaić V. Choise and planing of technology of construction of "Arena" in Zagreb related to meeting the deadline, Management tourism and sport, Sveti Martin na Muri. 2014

[13] Križaić V, Rodiger T, Buč S. Informa tizirani menadžer u vektorskoj organizac iji. Journal of Civil Engineering and Architecture. 2021;**15**(8):419-428. DOI: 10.17265/1934-7359/2021.08.003

[14] Križaić V, Završki I. Modelska standardizacija građevinske ponude metodom dinamičkog strukturnog programiranja, 5. Simpozij doktorskog studija građevinarstva, 9–10 Sep 2019 Vol. 117. Zagreb: University of Zagreb - BF; 2019. DOI: 10.5592/CO/PhDSym.2019.10

[15] Križaić V. Planning through combinatorics, VII International conference Organization, technology and management in construction. Zadar. 2006

[16] Križaić V, Hranj D. Project Planning by Modified Gauss S-Curve , CCC 2020, 28 June-1 July 2020, Opatija, Croatia, 2020. DOI: 10.3311/CCC2020-047

[17] Križaić V, Rodiger T, Baksa S. Simula tion Project Management by Modified Gaussian S-Curve, 14th WCCM-ECCOMAS Congress, January 2021. 2021 DOI:10.23967/wccm-eccomas.2020.309

[18] Fallera C, Höftmanna M. Service-oriented communication model for cyber-physical-production-systems, 11th CIRP Conference on Intelligent Computation in Manufacturing Engineering. Procedia CIRP. 2018;**67**:156-161

[19] Bonci A, Carbonari A, Cucchiarelli A, Messi L, Vaccarini M.

A *cyber*-physical system approach for b uilding efficiency monitoring. International Research Journal Automa tion in Construction. 2019;**102**(6):68-85. DOI: 10.1016/j.autcon.2019.02.010

[20] Monostori L, Kádár B, Bauernhansl T, Kondoh S, Kumara S, Reinhart G, et al. Cyber-physical systems in manufacturing. CIRP Annals – Manufacturing Technology. 2016;**65**(2): 621-641. DOI: 10.1016/j.cirp.2016. 06.005

[21] Madubuike OC, Anumba CJ. Potential for the Integration of Cyber-Physical Systems with Intelligent Buildings. ResearchGate. 2020. DOI: 10.1061/9780784482865.074

[22] Križaić V, Križaić N. Vectorial analysis of robots for hydrodemolying bridges. WMVC 2022 - 10th International Conference on Wave Mechanics and Vibrations/Dimitrovova, Z. (ur.). Lisbon: NOVA University of Lisbon; 2022. pp. 10-10

[23] Picek I. Elementarne čestice – iskrenje u svemiru tamne tvari, Školska knjiga, Zagreb. 1997

[24] Križaić V, Hranj D, Rodriger T. Utjecaj softverske matematike na modeliranje u graditeljstvu, 5. međunarodna konferencija Inovacije, tehnologije, edukacija i menadžment, 4. i 5. travnja 2018., Sveti Martin na Muri, Croatia. 2018

[25] Horvat I. Connectors of smart design and smart system. Artificial Intelligence for Engineering Design Analysis and Manufacturing. 2021;**35**:132-150. DOI: 10.1017/S0890060421000068

[26] Naticchia B, Messi L, Pirani M, Bonci A, Carbonari A, Tolve LC. Holonic System for Real-Time Emergency management in Buildings. ResearchGate. 2018. Available from: https://www.researchgate.net/publication/333894832

[27] Nogel M, Kovach G, Wersenyi G. The Regulation of Digital Reality in Nutshell. ResearchGate. 2022. Available from: https://www.researchgate.net/publication/354689836

[28] Križaić V, Hranj D, Blažon I. Građe vinski rizik" modificiranom Gaussovom S- krivuljom. OTMC 2022 Dubrovni/ Završki I, Cerić A, Vukomanović M, Huemann M, Golomejić Vlahov RA, Sigmund Z (ur.). Cavtat, Dubrovnik: HUOG, 2022. pp. 45-45

[29] Križaić V. Održivost mostova i nadvožnjaka zaobilaznice autoputa oko Zagreba, ISC People, Buildings and ldings and environment (PBE2014), Kromerič, Czech Republic. 2014

Chapter 5

IoT Framework Application to SHM Systems

Eduardo Hidalgo-Fort, Pedro Blanco-Carmona, Alvaro Serrano-Chacon, Emilio José Mascort-Albea, Fernando Muñoz-Chavero, Ramón González-Carvajal, Antonio Jesús Torralba-Silgado and Antonio Jaramillo-Morilla

Abstract

This chapter presents and details a complete IoT solution for secure, synchronous, low-power, wireless and unattended structural health monitoring (SHM), whose architecture can be parameterised to any type of structure to be monitored, thanks to its modular design based on an embedded real-time operating system at the node level and a microservices architecture at the application level. This solution is validated by means of its deployment, together with two commercial reference systems, in two experimental use cases in two different structures: on the one hand, a bridge located at the Engineering School of the University of Seville (Spain) and, on the other hand, the Homage Tower of the Utrera's Castle (Spain). The results obtained place this work as a novel solution in the current state of the art.

Keywords: Internet of Things (IoT), structural health monitoring (SHM), wireless sensor network (WSN), nondestructive test (NDT), low-cost SHM, IoT applications

1. Introduction

Over the last few decades, the relationship between humans and the environment surrounding us, as well as with the different frameworks, has changed considerably due to the emergence of new technologies that have made it possible increasing the quantity and heterogeneity of the information acquired from the environment, as well as from the manufacturing processes. However, this has resulted in a variety of non-standardised, and therefore non-compatible technologies and architectures that do not allow information exchange between them and limit the impact of these advances on the global state of the art. It is precisely this heterogeneous framework of needs and available technologies that gives rise to the emergence of the IoT (Internet of Things) paradigm.

According to IBM [1], IoT is the concept of interconnecting any device, thing, or person (things) to the Internet and to other devices and people to share data on their behaviour and that of the surrounding environment. Oracle [2] specifies how these ' things' can range from everyday objects to sophisticated industrial tools. This global integration is produced thanks to the definition of a standardised layered architecture [3, 4] shown in **Figure 1** and has a direct impact on the technology and its use, offering new capabilities and advantages over the previous framework, such as very low-cost and low-power devices with Internet connectivity and their integration in cloud processing platforms where they can host from simple applications to complex analysis processes based on machine learning techniques. The ability to integrate connected networked devices into a single system is one of the main features that make IoT a very attractive technology in the era of digital transformation.

The evolution of wireless communication technologies, the standardisation of IoT architecture and the globalisation of manufacturing processes invite the implementation of the new paradigm in these production lines, which has given rise to the concept of Industry 4. 0, the fourth industrial revolution characterised by the integration of advanced technologies to create smart factories, cyber-physical systems and IIoT (Industrial Internet of Things) [5]. The implementation of IoT in industry is becoming more and more common due to its numerous advantages, generating several industry-beneficial applications: (1) Predictive analytics and preventive maintenance improving by enabling data collection and subsequent analysis and helping predict machine failures and perform preventive maintenance before they occur, which in turn reduces costs and downtime [6], (2) Improving product quality by monitoring and controlling critical factors during the production process allowed for greater control over processes and reduction of defects [7], (3) Edge computing [8], technology responsible for processing data in real time at the point, where it is generated rather than sending it to a central location for processing, reducing processing time and load on central networks and servers and (4) Digital twin generation [9], enabling simulation and testing of different scenarios virtually, which reduces the costs and risks associated with testing and validation processes.

All these technological advances are reflected in the 5.5 billion devices connected in 2022 to operated networks (2G/3G, NB-IoT/CatM, 4/5G) according to the manufacturer Ericsson [10] on the 43 billion connected devices estimated for 2023

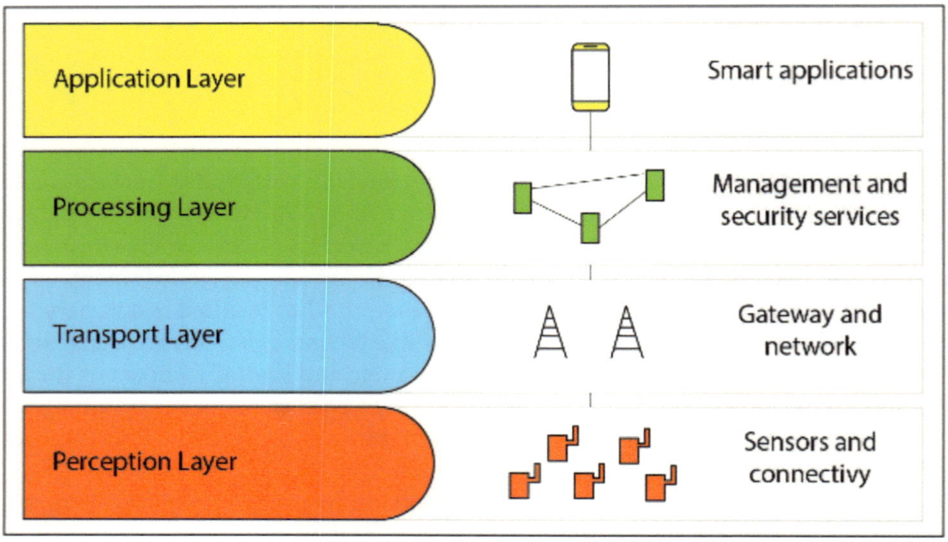

Figure 1.
Standardised IoT four-layer architecture.

by Forbes magazine [11]. On the other hand, this impact translates directly to the market, as predicted by the market analysis organisation Strait Research in its report [12], where it estimates a market evolution from USD 813 billion in 2021 to more than USD 2483 billion in 2030.

Thanks to the IoT benefits, it is applicable in different sectors such as smart agriculture [13–15], where more economically efficient ecosystems are generated by automating tasks such as cultivation and its monitoring allowing early detection of pests and diseases, in Healthcare [16–18], where remote and real-time monitoring, especially after the global pandemic of COVID-19, enables early and accurate detection of pathologies, while the application layer allows direct integration of patients with all links in the healthcare chain, resulting in new care models that improve the health and experience of patients, as well as the performance of healthcare professionals and smart homes [19, 20], where the integration of domestic elements in a secure and real-time network allows the optimization of different aspects such as energy management (even generating energy communities), the improvement of the feeling of comfort and the adaptation of the environment to the emotional state of the inhabitants. This framework has given rise to the smart city concept [21], which integrates structural monitoring of buildings, traffic control, vehicle-to-infrastructure (V2I) communications, waste management, noise monitoring or air quality, among others; and the automotive sector [22, 23], where the emergence of 5G communications allows real-time monitoring of vehicles and cloud computing the generation of digital twins [24] application of machine learning algorithms for vehicle and fleet management tasks from behavioural analysis, accident detection [25] and predictive maintenance.

These applications provide an overview of how SHM calculation could be automated with current technology. Today, using the range of components that comprise the IoT architecture, data can be obtained in an automated way, and, in real time, relevant indicators for determining structural health.

2. Structural health monitoring

Once a given structure has been built, whether it is a building or civil infrastructure, it is necessary to identify any damage that may compromise structural safety at an early stage. In this context, the structural health monitoring (SHM) plays a key role. According to the International Society for Structural Health Monitoring of Intelligent Infrastructures, it can be defined as: '*a type of system that provides on-demand information about any significant change or damage to the structure*' [26].

This discipline has experienced great growth in recent decades. Equipping, a structure with a structural monitoring system, has many advantages. A direct consequence of this is better control of structural safety, detecting in advance situations that could lead to collapse. Moreover, a proper structural diagnosis is vital to design an adequate repair or reinforcement intervention and, consequently, extend the useful life of the construction.

Basically, a structural health monitoring system consists of several elements, which can be grouped into six blocks [27]: (i) a sensory system (SS), (ii) a data acquisition and transmission system (DATS), (iii) a data processing and control system (DPCS), (iv) a data management system (DMS), (v) a structural health evaluation system (SHES) and (vi) an inspection and maintenance system (IMS). The first two blocks are located in the structure itself, while the next three blocks are normally

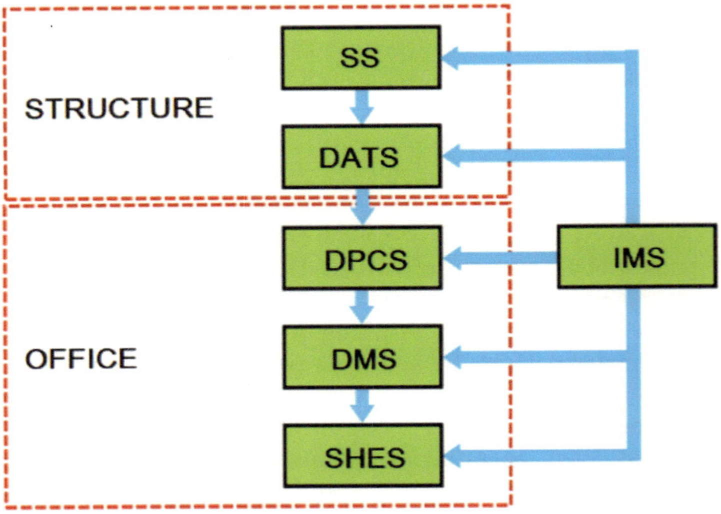

Figure 2.
Composition of a SHM system. Source: Author, based on Ref. [27].

located outside the structure. The last block supports all the others. The relationship between the listed blocks is graphically depicted in **Figure 2**.

Depending on the number of variables under study, the SHM system can be made up of a wide variety of sensors. In some cases, the focus is on taking data relating to environmental conditions, such as humidity or temperature. In other cases, the interest extends to data related to the response of the structure. This response can be expressed in terms of displacements, velocities or accelerations. These three variables have a very small magnitude when the structure is subjected to environmental excitations. Among them, acceleration is the most commonly used parameter in the dynamic characterisation of the construction by means of operational modal analysis.

The sensor used to measure the rate of change of velocity with respect to time is known as an accelerometer. Currently, these sensors can be classified according to the technology used, with the following types being distinguished: piezoelectric, piezoresistive, capacitive, balanced force, optical and thermal, among others. In each category, there are multiple variants depending on the number of axes, the range, the bandwidth or the price.

Traditionally, the collection of acceleration readings in the structure is performed with commercial accelerographs [28, 29]. This equipment is characterised by high accuracy, which usually involves high costs. Accelerograph prototypes developed around micro-electro-mechanical systems (MEMS) accelerometers are becoming increasingly common [30, 31]. In this way, it is possible to obtain devices that are very economical and at the same time very low in energy consumption [32]. The equipment constructed in this way is also lightweight and very small in size, which makes it easy to transport and place in structures.

Acceleration data collected with accelerometers are used for the estimation of the dynamic properties of structures. For this purpose, a large number of well-known operational modal analysis methods are available. These methods are usually classified according to the domain in which the identification of modal parameters takes place, with two groups being distinguished: (i) time domain methods, which operate directly on the signals recorded in tests or on correlation functions, such as natural excitation technique [33], covariance-driven stochastic subspace identification [34] and data-driven stochastic subspace identification [35] and (ii) frequency domain methods, which require the calculation of spectral density functions. This second group includes, among others, the following methods: maximum likelihood [36], frequency domain decomposition [37] and enhanced frequency domain decomposition [38].

The complete dynamic characterisation of a structure requires the determination of the three basic dynamic properties of the structure, namely natural frequencies, modal displacements and modal damping factors. When the structure deteriorates, either due to the passage of time or a sudden event such as an earthquake or a vehicle collision, a variation of these dynamic properties occurs. Therefore, in order to assess the deterioration of the building, it is necessary to compare two states: (i) initial or reference, which generally corresponds to a situation in which the structure has not been damaged and (ii) final, in which it is expected to have suffered some type of damage.

Perhaps the simplest way to detect damage in a construction is by the reduction of the natural frequencies of the structure. This is evidenced by the large number of scientific papers published in the last quarter of a century related to structural diagnostics. Under the assumption that the mass remains constant, the progression of deterioration will lead to a decrease in the stiffness of the structure. Consequently, there will be a reduction of the natural frequencies. However, the observation of significant changes in the natural frequencies requires a considerable deterioration of the structure. In the other direction, an increase in this dynamic property is useful to check the effectiveness of certain repair or reinforcement actions on the structure under study.

Typically, structural damage is classified into four levels according to the degree of information that is available about it. In this context, it is common practice in the scientific community to use the damage levels proposed by Rytter [39], which are summarised in **Table 1**.

In terms of the level of damage, methods based on changes in natural frequencies are only useful to indicate the presence of damage to the structure (Level I). When a higher degree of information on the deterioration of the building is required, techniques using variations in modal displacements are often used. In this way, it is possible to locate the most likely area in which the damage is located (Level II).

On the other hand, civil engineering has not been transparent to the technological evolution prior to the emergence of the IoT and, as a consequence, numerous solutions have appeared for different fields of application [40–42], which have not been able to mature enough to become widely adopted solutions by the community, mainly due to the changing and constant emergence of new technologies and, therefore, have

Level	Name	Description
Level 1	Detection	Indication of the presence of damage to the structure
Level 2	Localisation	Level I + Geometric location of damage
Level 3	Assessment	Level II + Estimation of the extent of damage
Level 4	Prediction	Level III + Structural safety prediction

Source: [39].

Table 1.
Levels of damage on structures.

not generated systems for the extraction of advanced information from structures (Level III or Level IV).

However, the emergence of the IoT paradigm and its standardised architecture as an integrative ecosystem eliminates dependencies between layers, and thus allows for expert contributions from each knowledge area to its corresponding layer. In other words, this allows the development and evolution of each of the functional blocks of a SHM system (data processing, hardware design, firmware design, high-level applications, etc.) independently in a framework that ensures the integration of all of them and the bridging of desired levels III and IV.

3. Proposed solution

Once the application framework has been analysed, the design of the proposed SHM-IoT solution is carried out. To do this, initially, an analysis of the main requirements to be met by the resulting design will be carried out and, subsequently, the most important implementation aspects that give rise to the solution will be detailed.

3.1 Requirements

Based on the usual IoT device architecture model, the proposed solution is composed of four functional layers (**Figure 1**) with the following scope:

- Perception layer: These are the devices or nodes in charge of carrying out the monitoring of the structure under analysis by acquiring the magnitudes of interest.

- Transport layer: This is the infrastructure responsible for guaranteeing a communication channel between the monitoring nodes and the processing and application layers.

- Processing layer: It is in charge of housing the data center with the capacity to execute data processing algorithms (traditional or machine learning) and offer its services to the application layer.

- Application layer: Together with the processing layer, it makes up the application server (physical or virtual) and hosts the applications necessary to carry out both the exploitation of the information generated by the processing layer and the management of the monitored infrastructures.

In this way, the integration of the solution in the IoT paradigm is guaranteed. The specific requirements of each of the proposed layers are detailed below.

3.1.1 Perception layer

The measurement devices are one of the most important elements of the proposed solution as they are responsible for monitoring the behaviour of the structure under test and the surrounding environment. Ultimately, therefore, they are the interface between the rest of the system and the infrastructure. In terms of technological requirements and challenges, the following are the most important.

3.1.1.1 Versatility and scalability

From a functional point of view, for frequency analysis, it is necessary for the nodes to have the capacity to monitor structural accelerations. Nevertheless, each structure has different monitoring needs depending on its architecture, typology or impact on society. For this reason, the proposed solution must be versatile in terms of the type of magnitudes to be monitored, which would allow it to be adapted to any civil infrastructure.

In the same way, each structure will have different needs in terms of the number of monitoring points, so the scalability of the solution becomes one of the enabling factors to ensure its success.

3.1.1.2 Wireless connection

Even though the IoT framework allows for the integration of both wireless and wired devices, the proposed application involves the monitoring of civil, strategic and historical heritage infrastructure, where the minimisation of visual pollution is an important factor. This implies devices as small as possible. On the other hand, from an economic point of view, the elimination of the high cost of cabling and its installation and maintenance, added to the low cost of wireless communications, makes wireless communications a must.

3.1.1.3 Time synchronisation

Frequency analysis of structures implies the need for time-synchronised devices so that the structural behaviour of the monitored building can be reconstructed for subsequent analysis using operational modal analysis (OMA) techniques.

3.1.1.4 Low energy consumption

One of the key aspects for the success of the solution is to meet very strict low-power requirements, so as to guarantee the operability of the deployed solution without the need for an external power supply (many infrastructures do not have an electrical installation) or to carry out maintenance work on it. These tasks would include the replacement of batteries, which has a direct impact on the cost of the solution and avoids the management of access to the infrastructures, which presents an added difficulty in the case of structures belonging to historical heritage.

3.1.1.5 Remote updating

Since this is a novel solution and a line of research in constant evolution, added to the fact that the structural particularity of each construction implies a parameterisation of the monitoring systems (full scale, sampling frequency, preprocessing, etc.), the need to carry out firmware updates of the devices deployed is guaranteed. Therefore, it is necessary for the measurement nodes to have FOTA (Firmware over the air) functionality that allows them to be updated without the need for physical action (usually located in areas that are difficult to access), avoiding interaction with strategic structures or historical heritage, as well as the management of complex bureaucratic permits and helping to reduce maintenance costs.

3.1.2 Transport layer

Another of the key aspects for the implementation of the proposed solution is the communications infrastructure used to send information between its different functional elements.

On this point, it is important to note that the solution presented must be suitable for monitoring a structure or a set of structures that may be located in one geographical point or distributed throughout the entire regional, autonomous or national geography (depending on the organisation that operates them), which, due to costs, rules out the option of using local networks maintained by the owners of the structures and highlights the need for a solution that offers communications to all of them globally, which makes it necessary to think of operated networks.

On the other hand, given that it may be a strategic infrastructure, it is necessary that the communications network offers sufficient security mechanisms to guarantee the integration and privacy of the information transmitted, as well as long-term operability.

3.1.3 Processing layer

The third main element is the application server, which will be in charge of providing secure services to both the processing layer and the application layer.

While integration in an IoT framework involves hosting services on a server (physical or virtual) in the cloud, it is necessary to think of a design based on microservices that allow scaling the functionalities of the deployed solution without the need to redesign the previous deployment and, therefore, will facilitate the integration of new value-added applications. This feature will also allow the migration of deployed services with minimum effort from one service provider to another if necessary.

On the other hand, the large amount and heterogeneous nature of data collected from infrastructures require an architecture and engine capable of integrating, moving and processing such data. For this reason, a non-relational database architecture and big data should be chosen.

The processing layer will oversee housing the processing algorithms in charge of extracting the structural characteristics of the monitored buildings from the data collected by the perception layer. Finally, these results will be offered to the application layer for subsequent exploitation.

3.1.4 Application layer

Finally, like the processing layer and the application layer should be composed of those microservices necessary to ensure the management of the monitored assets and provide the end user with visualisation of results and secure access based on roles and permission schemes.

3.2 IoT solution implementation details

3.2.1 Perception layer

From a hardware design point of view, the monitoring nodes have an architecture of two printed circuit boards (PCB) with different functionality. On the one hand, the core board is responsible for the transversal functionalities of any infrastructure to be monitored such as power supply, local processing and control of the node (STM32L152 microprocessor), local storage of information (1024Kb EEPROM memory and MicroSD card) and management of wireless communications (NB-IoT-SIM7080G). On the other hand, the sensor board is responsible for housing the necessary sensorization in each structure under analysis. Although the fundamental elements will be the accelerometer (ADXL355) and the GPS module (A2235H) that will guarantee the synchronisation of all the elements of the perception layer, this board has analogue and digital interfaces (SPI, IIC and UART) for the connection of other useful sensors depending on the type of structure (strain gauges, temperature and humidity, pressure, etc.). Finally, both boards are housed in an IP68 enclosure together with a PCB antenna for NB-IoT communication.

From a firmware design point of view, the need of time synchronisation and the parallelisation of the tasks of measurement acquisition, local storage, data processing and information sending have implied the use of a real-time operating system, opting for FreeRTOS. The parallelization of the tasks of the monitoring nodes and the use of the low consumption modes of the microprocessor and the sensors, as well as the efficient management of the microSD card, have allowed the optimisation of the energy consumption of the devices, maximising the autonomy of the device up to 2 years and 4 months with a 17 Ah battery capacity [32].

Additionally, the devices have firmware over the air (FOTA) functionality as another task of the node, which allows, after a connection is initiated by the node, checking for the existence of new firmware version and, if necessary, updating its functionality remotely and wirelessly, without the need to access them physically, guaranteeing the maintenance of the deployed solution.

3.2.2 Transport layer

Considering the requirements defined for the transport layer, the NB-IoT-operated network offers a level of coverage [43] and sufficient operators [44] to guarantee a service that satisfies them. Moreover, the market prices [45] guarantee the cost-effectiveness of the service compared to the deployment of proprietary local networks, such as LORA, which require the maintenance of the owner and a gateway to the operated network.

Another important aspect of SHM is to guarantee the integrity and security of the transmitted information, which may be sensitive information pertaining to strategic structures or the historical heritage of a city, region or country. For this reason, a secure communications stack is proposed, consisting of a transmission control protocol (TCP) connection-oriented communication service that guarantees that the transmitted information always reaches its destination and a message queuing telemetry transport (MQTT) service that minimises the number of established connections (light service), offers different QoS (Quality of Service) levels and guarantees the integration of the information in the IoT services offered by all the reference operators (Amazon WebServices, Azure, etc.). In addition, the selection of MQTT allows the use of technologies, such as SSL/TLS and user authentication *via* password or certificate.

Therefore, by opting for NB-IoT as the communication technology and the TCP/MQTT/SSL stack, we have a service that satisfies all the proposed security requirements.

3.2.3 Processing and application layers

As discussed in the requirements for these layers in Section 4.1, its implementation as a microservices deployment is a fundamental aspect. In this sense, Docker is selected as the container technology because it is the most mature of the solutions offered in the current catalogue.

The microservices that make up the solution at server level are the following:

- Reverse proxy: This is a service for port forwarding that enables communication between the different microservices. Nginx technology is chosen as it is accessible from the outside with HTTPS protocol, protected by the use of certificates.

- Decoder: Programmed in Python, this is the service in charge of receiving the information collected by the perception layer and interpreting its information before storing it in the database. It is also responsible for encoding the information downlinked to the nodes.

- Backend: This is the server core. It uses node.js image and it is in charge of the management and maintenance tasks of the processing and application layers, as well as being the interconnection with the frontend.

- Data processing (Processing layer): This is the data processing service. Developed in Python, it hosts the information processing algorithms, in this case, the detection of fundamental frequencies, and has the capacity to host the processing and machine learning services that may be considered in the future.

- Redis DB (Processing layer): Service whose function is purely the intercommunication of the backend with the data processing module, communication necessary to receive orders for training and updating behavioural models (deterministics or machine learning models) from the frontend and backend, and to transmit these orders to the Data processing module.

- Mongo DB: This is a non-relational database data storage service. This type of database engine allows the hosting of heterogeneous data that does not respond to a fixed structural pattern (different kinds of structures and different sensors), as occurs in relational databases. In this case, a Mongo 5.0.9 image is used and allows the creation of a big data architecture, as desired.

- Frontend: It is the service for the client side, and it uses a node.js 18.6.0 image. It is in this service where the web interface will be mounted to which the user will have access and where they will be able to consult the data received, manage the set of monitored infrastructures, as well as the nodes deployed in the field and their firmware updates.

Therefore, the proposed design results in a versatile and complete IoT solution for SHM, which will be validated in the use cases proposed in section V.

4. Case studies

Currently, there is a great deal of concern among researchers about historic buildings as they have a high patrimonial value. In this sense, it is worth highlighting the research carried out by Zini et al. [46], in which the 'Torre Grossa' (San Gimignano, Italy) and the Mastio di Matilde (Livorno, Italy) were studied. In order to obtain a more complete structural diagnosis, it is common not only to take sensor data but also to carry out visual inspections of the structure. This procedure was successfully applied in the Zuccaro's Tower, a medieval masonry building located in the Italian city of Mantua [47]. The execution of environmental vibration tests can also be applied to ancient bridges. As an example of this, it is worth noting the study of the Turkish bridge spanning the Firtina Creek [48] and the Roman bridge near the Italian town of Todi [49].

In other cases, the identification of the dynamic properties of these historical structures experimentally is intended for the calibration of finite element models. That is, the adjustment of the model parameters to achieve a higher degree of approximation to the real behaviour of the structure. In this way, the resulting model allows for a more rigorous assessment of future scenarios, such as earthquakes or ground settlements. In the field of historic buildings, it is worth mentioning the work carried out by Venanzi et al. [50] on the Sciri Tower (Perugia, Italy) and the study carried out by Standoli et al. [51] on the Civic Tower (Ostra, Italy). With regard to historic civil infrastructures, it is interesting to mention the research carried out on the St. Lázaro and the Lagoncinha bridges [52] and the Durrães bridge [53], all of them located in Portugal.

Structural health monitoring is also of interest for more recent constructions, especially those that are considered to be unique or vital for the society. The common denominator of these buildings is the use of contemporary materials, such as concrete or steel. In this context, without being exhaustive, the following examples can be highlighted: the Infante D. Henrique Bridge, a concrete arch bridge in Portugal [54]; a steel cable-stayed footbridge over the Hron River in Slovakia [55]; the roof of the Municipal Sports Stadium of Braga, also in Portugal [56]; the Morrow Point Dam in the United States [57]; a tall building in Qatar [58].

As it can be seen, there is a wide variety of structures that are equipped with a structural health monitoring system. The results obtained are extremely useful, both to protect the lives of users and to properly manage the resources allocated to the conservation of the structures.

The solution presented in this chapter is an IoT system, which has all the components of a structural health monitoring system. The following subsections focus on node validation. For this purpose, two case studies located in Spain have been selected, namely: the Homage Tower of the Utrera's Castle and a bridge close to the Engineering School of the University of Seville. These two structures are completely different in their characteristics, which is intended to demonstrate the versatility of the presented node. Specifically, the first case study represents a historical masonry building with a predominant vertical dimension and no structural code was used in its design. However, the other case study consists of a more recent concrete structure, which has the largest dimension in the horizontal plane and follows strict design criteria.

4.1 Methodology

In both case studies, the same procedure is used for the experimental determination of the dynamic properties of the structures. In particular, the focus is on the determination of the natural frequencies.

The process for the dynamic characterisation of these structures consists of two phases. In the first of these, a set of environmental vibration tests is carried out. For this purpose, natural sources of excitation, such as wind or traffic, are used. These sources of excitation cause the structure to vibrate. The structural response in terms of accelerations is recorded using accelerographs. This data collection campaign is carried out simultaneously with the node presented in this chapter and commercial equipment.

In each environmental vibration test, acceleration readings are taken at two points on the structure. One of these points is common to all the tests and serves as a reference. Proceeding in this way, in each test there are a total of six series of accelerations since the accelerometers are triaxial. Ideally, it would be desirable to have as many accelerographs as measurement points on the structure. This significantly reduces the duration of the experimental campaign but is unfeasible due to the high cost of the commercial measuring equipment. However, it would be feasible with the node presented due to its low cost.

In the second phase, operational modal analysis methods are applied in both the time and frequency domain. This allows the identification of the dynamic properties of the structures. The commercial software ARTeMIS Modal Pro is used to carry out this task, which is widely used in the scientific community.

4.2 Expo-92 bridge

This structure is located a few metres from the Engineering School of the University of Seville and, specifically, in the Water Street. It consists of a reinforced concrete bridge with a curved direction, with a length of 38 metres. The deck is 16.20 metres wide, which allows for the passage of both people and vehicles. The deck is supported by the abutments and two groups of isolated piers, which are also made of reinforced concrete. The piers have a hexagonal cross section and are separated by a distance of 4.80 m in the transverse direction of the bridge. A panoramic view of the bridge under study is shown in **Figure 3**.

4.2.1 Dynamic characteirsation

According to **Figure 4**, data collection on the bridge is carried out at 28 points, distributed in four rows in the longitudinal direction of the bridge. The accelerometers are placed on the bridge deck, with the z-axis in the direction of gravity and upwards. Point P1 was common to all ambient vibration tests. In order to measure all the points of the structure, eight tests were carried out with the commercial equipment and eight tests with the presented node.

As a reference, eight 352C33 accelerometers from the manufacturer PCB piezotronics were used. The output signals from the sensors are collected by an eight-channel data acquisition system (DAQ) and, in particular, the SIRIUS-8xACC model from the manufacturer Dewesoft. The transformation of the analogue to digital signal takes place by means of a 24-bit ADC embedded in the DAQ. For this equipment, a measuring range of ±2 g was set, as with the node presented. A sampling frequency

Figure 3.
Expo-92 bridge. Seville, Spain.

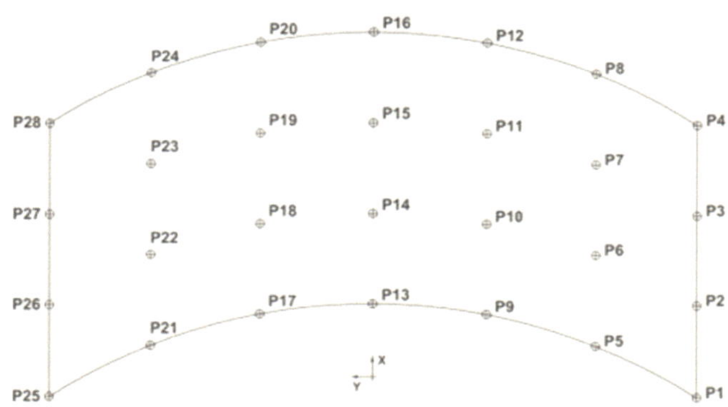

Figure 4.
Measurement points on the bridge.

of 100 Hz was adopted for the commercial equipment. In contrast, a value of 31.25 Hz was adopted for the low-cost equipment since the interest frequencies are below 10 Hz and this sampling frequency satisfies Nyquist theorem. Prior to the experimental campaign, a finite element model was prepared. This preliminary study made it possible to define the duration of the time series with a better criterion, establishing a length of 8 minutes. **Figure 5a** illustrates the collection of data with the commercial equipment, while **Figure 5b** shows the low-cost equipment.

As mentioned above, the identification of the natural frequencies of the bridge is done with ARTeMIS Modal Pro. Only the bridge deck is introduced in this software. The 28 points measured on the structure define a grid, with all points at the same elevation. The x and y coordinates of the points match **Figure 4**. By proceeding in this way, the model shown in **Figure 6** is obtained. The acceleration series recorded during the tests are assigned to each of these nodes.

The dynamic characterisation of the bridge under study has been carried out using an operational modal analysis method in the frequency domain and another in the time domain. In particular, the EFDD and SSI-UPC-Merged methods are used, obtaining practically the same results. The first four modes of vibration of the

Figure 5.
Ambient vibration test on the bridge: (a) 352C33 accelerometer and (b) node.

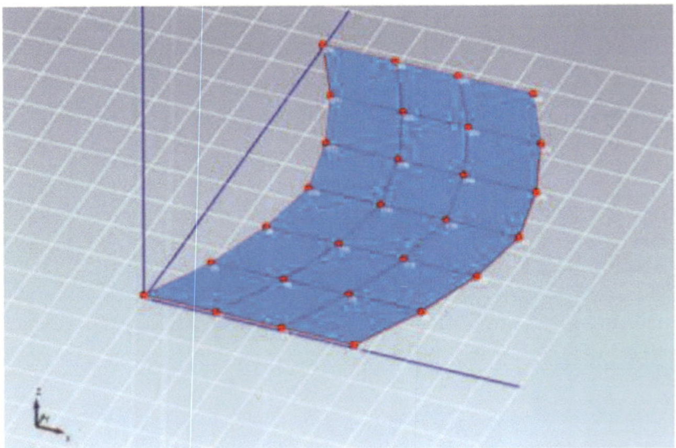

Figure 6.
Bridge geometry in ARTeMIS modal pro.

structure are located in the range between 0 Hz and 10 Hz. A summary of the natural frequencies obtained with the EFDD method, both with commercial and low-cost equipment, is given in **Table 2**. The maximum relative error in the identification of this modal parameter with the equipment presented here is 9.68%.

The low-cost equipment is able to estimate the second natural frequency of the bridge with an excellent degree of approximation, with a relative deviation of less

Mode	Campaign N° 1—352C33 (Hz)	Campaign N° 2—This work (Hz)	Deviation (%)
Mode 1	2.808	2.960	5.41
Mode 2	5.493	5.439	0.98
Mode 3	7.448	7.075	5.01
Mode 4	8.060	8.840	9.68

Table 2.
Dynamic characterisation of the bridge.

than 1%. In the identification of the fundamental frequency and the frequency associated with the third mode of vibration, the relative error increases to about 5%. This difference is even greater in the last detected vibration mode. It should be noted that the equipment presented has a significantly lower cost than commercial equipment but, even so, it allows the identification of the first three modes of vibration of the structure with a good degree of approximation. Although the deviation is higher in other modes of vibration, this is not very relevant as the higher modes are less significant from a structural point of view.

4.3 The homage tower of the Utrera's castle

This building belongs to the Utrera's Castle and is located in the town of the same name in Seville. The castle is considered to be built between the thirteenth and fourteenth centuries. This justifies the use of older building materials such as rammed earth, brick and stone. The tower under study has a rectangular cross section, with a smaller side of 11.50 m and a larger side of 12.12 m. The distance between the base and the parapet of the roof is 18.55 m. The walls are 1.90 m thick, although the side with the stairs is thicker. The tower is divided in height into two chambers, which are limited at the top by two vaults. In **Figure 7**, it can be seen a photograph of the current state of the Homage Tower.

4.3.1 Dynamic characterisation

The recording of accelerations in the tower is carried out at eight points, which are shown in **Figure 8**. In this last figure, the orientation of the accelerographs can be seen, with the z-axis in the direction of the gravitational acceleration, but in an upward direction. Point A was taken as the reference point for all tests so that seven ambient vibration tests were carried out at the Homage Tower.

Figure 7.
Homage tower of the Utrera's castle.

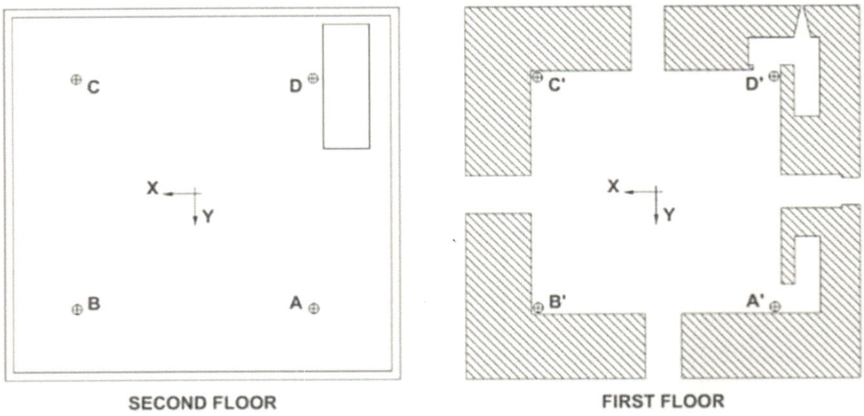

Figure 8.
Measurement points on the tower.

The first experimental campaign was done with two GMS Plus6-73 devices, manufactured by GeoSIG. This device contains, among other components, a triaxial accelerometer, a 24-bit analogue-to-digital converter, a GPS and a battery. The second experimental campaign was carried out with the low-cost equipment. In both cases, a measurement range of ±2 g was taken. However, there are differences with regard to the sampling frequency. In the GMS Plus6-73 station, a value of 200 Hz was used, while in the developed node, a value of 125 Hz was adopted. Each test has a duration of 10 minutes. **Figure 9** shows photographs taken during the execution of the tests with the two devices.

The acceleration series recorded during the tests are transferred to the ARTeMIS Modal Pro. Previously, it is necessary to generate the geometry of the structure in a simplified form. This is composed of the eight points, where readings have been taken. In addition, the base of the tower is represented by four points. These have the same x and y coordinates as the points of the first floor but are located at zero elevation. The model generated in this way can be seen in **Figure 10**.

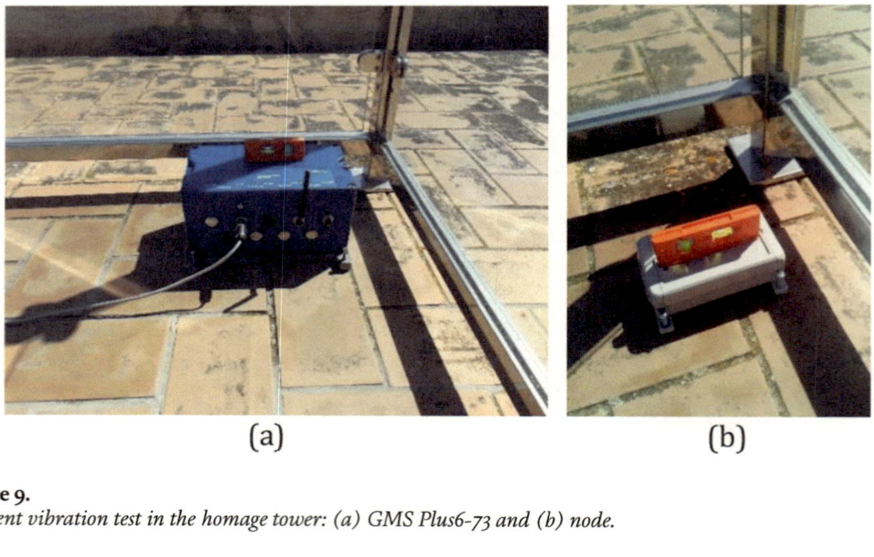

Figure 9.
Ambient vibration test in the homage tower: (a) GMS Plus6-73 and (b) node.

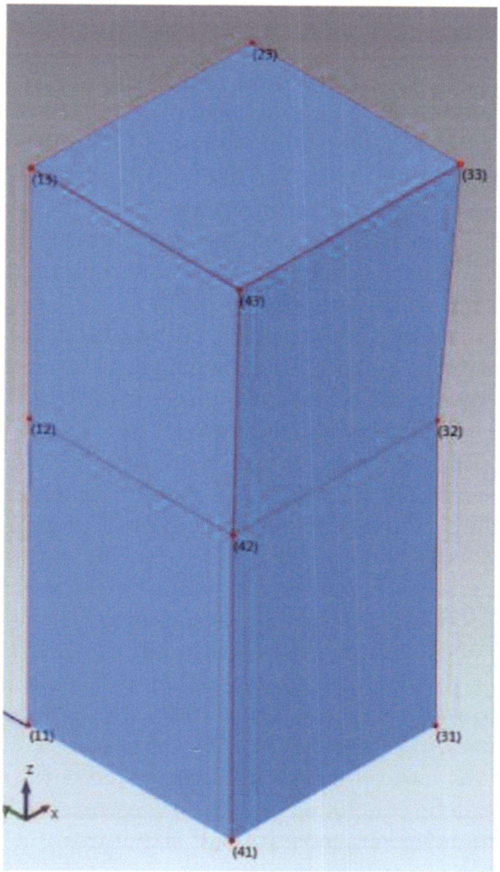

Figure 10.
Tower geometry in ARTeMIS modal pro.

The estimation of the natural frequencies of the Homage Tower is also carried out with the EFDD and SSI-UPC-Merged methods. The results obtained with the EFDD method in the two experimental campaigns are summarised in **Table 3**. In the range from 0 Hz to 10 Hz, two modes of vibration were identified. As it can be seen in this table, the maximum relative error between the natural frequencies obtained with both experimental campaigns is 0.08%.

The identification of natural frequencies with the reported node is less than 0.1% compared to the commercial equipment. This highlights the usefulness of low-cost equipment as a diagnostic tool in heritage buildings.

Mode	Campaign N° 1—GMS Plus6-73 (Hz)	Campaign N° 2—This work (Hz)	Deviation (%)
Mode 1	3.359	3.357	0.06
Mode 2	7.208	7.202	0.08

Table 3.
Dynamic characterisation of the tower.

5. Conclusions

This chapter introduces the IoT paradigm and its standardisation as a strategic framework for the development of large-scale deployable structural health monitoring systems. Following the presentation of the emergence and evolution of the IoT, as well as the opportunity of its application to the field of structural monitoring, this work has made it possible to reflect the importance of structural monitoring in the field of civil engineering and architecture. This relevance is even more important in the field of heritage and historical buildings, where construction techniques and technical solutions are not usually standardised.

Based on these premises, the requirements that would enable such a mass deployment of SHM solutions have been presented, which have resulted in a secure, modular and scalable architecture that allows its adaptability to different types of potentially monitorable structures, such as bridges and towers.

Subsequently, the proposed architecture has been validated through its development and subsequent real field tests (TRL7 level) such as the Expo92 bridge located at the Engineering School of the University of Seville (Spain) and the Homage Tower of the Utrera's Castle (Spain). Parallel monitoring with two traditional commercial reference systems has made it possible to validate the results obtained by the developed solution. With these data sets, the natural frequencies of the two structures were estimated. The error in the estimation of the fundamental frequency is around 5% in the first case study and below 0.1% in the second case study. With regard to the natural frequencies associated with the second and third modes of vibration, deviations within the range of 5% are observed. These deviations are more pronounced for the natural frequencies associated with the higher modes. With these results, it can be concluded that the node presented in this paper can be used to carry out a dynamic characterisation of structures and, at least, to detect damage on them.

From the point of view of signal processing, the architecture presented proposes a deployment of independent microservices that allow the application of different processing techniques depending on the monitored structure, which is presented as an opportunity, taking advantage of the trend of recent years, to integrate artificial intelligence and deep learning algorithms, such as [51] that allow Levels III and IV of SHM to be achieved. On the other hand, minimising energy consumption is always a technological challenge in any wireless monitoring solution such as the one presented, so minimising this energy consumption, mainly through Edge Computing techniques that minimise the amount of data to be transmitted wirelessly, is one of the most relevant current and future lines of research in this field of application.

The presented solution, which has been developed according to the stack of layers proposed by the IoT framework, allows centralised efforts to be applied in the different layers without altering the relationship between them, which guarantees the future integration of the different SHM solutions that may appear in the state of the art, both for signal processing in the application layer and in the perception layer, as well as for the migration to other types of wireless communication that may appear. Therefore, it can be concluded that this work presents and validates an IoT-SHM solution for low-cost, secure, scalable, low-power, wireless and large-scale deployable structural monitoring, which positions it as a novel reference in the current state of the art.

Acknowledgements

This work was fully supported by proyecto PID2019-107258RB-C31 financiado por MCIN/ AEI /10.13039/501100011033. The authors of this work also thank the Ayuntamiento de Utrera for their collaboration in the research.

Appendices and nomenclature

DAQ	data acquisition
DATS	data acquisition and transmission system
DMS	data management system
DPCS	data processing and control system
EFDD	enhanced frequency domain decomposition
FOTA	firmware over the air
IMS	inspection and maintenance system
IIC	inter-integrated circuit
IoT	Internet of Things
LoRaWAN	long range wide area network
LTE	long-term evolution
MEMS	micro-electro-mechanical systems
MQTT	message queuing telemetry transport
NB-IoT	narrow band internet of things
OMA	operational modal analysis
PCB	printed circuit board
RTOS	real-time operating system
QoS	quality of service
SHES	structural health evaluation system
SHM	structural health monitoring
SPI	serial peripheral interface
SS	sensory system
SSL	secure sockets layer
SSI-UPC	stochastic subspace identification-unweighted principal component
TCP	transmission control protocol
TLS	transport layer security
TRL	technology readiness level
UART	universal asynchronous receiver-transmitter

Author details

Eduardo Hidalgo-Fort[1*], Pedro Blanco-Carmona[1], Alvaro Serrano-Chacon[2], Emilio José Mascort-Albea[2], Fernando Muñoz-Chavero[1], Ramón González-Carvajal[1], Antonio Jesús Torralba-Silgado[1] and Antonio Jaramillo-Morilla[2]

1 Electronic Engineering Group, University of Seville, Seville, Spain

2 Department of Building Structures and Geotechnical Engineering, University of Seville, Seville, Spain

*Address all correspondence to: ehidalgo@us.es

IntechOpen

© 2024 The Author(s). Licensee IntechOpen. This chapter is distributed under the terms of the Creative Commons Attribution License (http://creativecommons.org/licenses/by/3.0), which permits unrestricted use, distribution, and reproduction in any medium, provided the original work is properly cited.

References

[1] Clark J. What Is the Internet of Things, and How Does it Work? IBM Blog. 2023. Available from: https://www.ibm.com/blog/what-is-the-iot

[2] What Is the Internet of Things (IoT)? (n.d.) Available from: https://www.oracle.com/internet-of-things/what-is-iot/

[3] Yaqoob I, Ahmed E, Hashem IaT, Ahmed AIA, Gani A, Imran M, et al. Internet of things architecture: Recent advances, taxonomy, requirements, and open challenges. IEEE Wireless Communications. 2017;**24**(3):10-16. DOI: 10.1109/mwc.2017.1600421

[4] Abosata N, Al-Rubaye S, Inalhan G, Emmanouilidis C. Internet of things for system integrity: A comprehensive survey on security, attacks and countermeasures for industrial applications. Sensors. 2021;**21**(11):3654. DOI: 10.3390/s21113654

[5] Chen B, Wan J, Shu L, Li P, Mukherjee M, Yin B. Smart factory of industry 4.0: Key technologies, application case, and challenges. IEEE Access. 2018;**6**:6505-6519. DOI: 10.1109/access.2017.2783682

[6] Him LC, Poh YW, Pheng LS. IoT-based Predictive Maintenance for Smart Manufacturing Systems; 2019. DOI: 10.1109/apsipaasc47483.2019.9023106

[7] Hu X, Zhana X. A Novel Method for Product Quality Control Based on Production IoT Data; 2022. DOI: 10.1109/iccbd56965.2022.10080209

[8] Kong X, Wu Y, Wang H, Xia F. Edge computing for internet of everything: A survey. IEEE Internet of Things Journal. 2022;**9**(23):23472-23485. DOI: 10.1109/jiot.2022.3200431

[9] Qi Q, Tao F. Digital twin and big data towards smart manufacturing and industry 4.0: 360 degree comparison. IEEE Access. 2018;**6**:3585-3593. DOI: 10.1109/access.2018.2793265

[10] IoT Connections Outlook. IoT connections outlook. In: Broadband IoT (4G/5G) Connections to Dominate by End of 2028. 2022. 2023. Available from: https://www.ericsson.com/en/reports-and-papers/mobility-report/dataforecasts/iot-connections-outlook?gclid=CjwKCAjwvdajBhBEEiwAeMh1U6Eh4JZgIjS9XqxAozf0iiwbEMTKWjqU04PPTU7S2SFS2c2v3NxvvhoCdVkQAvD_BwE&gclsrc=aw.ds

[11] Marr B. The Top 4 Internet of Things Trends In 2023. Forbes; 2022 Available from:https://www.forbes.com/sites/bernardmarr/2022/11/07/the-top-4-internet-of-things-trends-in-2023/

[12] Research S. Internet of Things (IoT) Outlook, Growth, Report to 2030. 2023. Available from: https://straitsresearch.com/report/internet-of-things-market

[13] Ayaz M, Ammad-Uddin M, Sharif Z, Mansour A, Aggoune EM. Internet-of-things (IoT)-based smart agriculture: Toward making the fields talk. IEEE Access. 2019;7:129551-129583. DOI: 10.1109/access.2019.2932609

[14] Qazi S, Khawaja BA, Farooq QU. IoT-equipped and AI-enabled next generation smart agriculture: A critical review, current challenges and future trends. IEEE Access. 2022;**10**:21219-21235. DOI: 10.1109/access.2022.3152544

[15] Elijah O, Rahman TA, Orikumhi I, Leow CY, Hindia MN. An overview

of internet of things (IoT) and data analytics in agriculture: Benefits and challenges. IEEE Internet of Things Journal. 2018;**5**(5):3758-3773. DOI: 10.1109/jiot.2018.2844296

[16] Bhuiyan MAS, Rahman M, Billah MM, Saha DK. Internet of things (IoT): A review of its enabling Technologies in Healthcare Applications, standards protocols, security, and market opportunities. IEEE Internet of Things Journal. 2021;**8**(13):10474-10498. DOI: 10.1109/jiot.2021.3062630

[17] Catarinucci L, De Donno D, Mainetti L, Palano L, Patrono L, Stefanizzi ML, et al. An IoT-aware architecture for smart healthcare systems. IEEE Internet of Things Journal. 2015;**2**(6):515-526. DOI: 10.1109/jiot.2015.2417684

[18] Kumar A, Krishnamurthi R, Nayyar A, Sharma K, Grover V, Hossain E. A novel smart healthcare design, simulation, and implementation using healthcare 4.0 processes. IEEE Access. 2020;**8**:118433-118471. DOI: 10.1109/access.2020.3004790

[19] Kelly SDT, Suryadevara NK, Mukhopadhyay SC. Towards the implementation of IoT for environmental condition monitoring in homes. IEEE Sensors Journal. 2013;**13**(10):3846-3853. DOI: 10.1109/jsen.2013.2263379

[20] Dorri A, Kanhere SS, Jurdak R, Gauravaram P. Blockchain for IoT Security and Privacy: The Case Study of a Smart Home; 2017. DOI: 10.1109/percomw.2017.7917634

[21] Zanella A, Bui N, Castellani A, Vangelista L, Zorzi M. Internet of things for smart cities. IEEE Internet of Things Journal. 2014;**1**(1):22-32. DOI: 10.1109/jiot.2014.2306328

[22] Chen QB, Schaeffer JK. 22FDX® Embracing IoT, 5G, and Automotive Applications—A Perspective through Global Research; 2019. DOI: 10.1109/s3s46989.2019.9320648

[23] Kunz A, Nkenyereye L, Song J. 5G evolution of cellular IoT for V2X. 2018. DOI: 10.1109/cscn.2018.8581830

[24] Fuller A, Fan Z, Day C, Barlow C. Digital twin: Enabling technologies, challenges and open research. IEEE Access. 2020;**8**:108952-108971. DOI: 10.1109/access.2020.2998358

[25] Kumar N, Acharya D, Lohani D. An IoT-based vehicle accident detection and classification system using sensor fusion. IEEE Internet of Things Journal. 2021;**8**(2):869-880. DOI: 10.1109/jiot.2020.3008896

[26] ISHMII. Definitions: Structural Health Monitoring. (n.d.) Available from https://ishmii.org/knowledge-education-center/education/shm-glossary/definitions/#S [Accessed: December 21, 2022]

[27] Ni YQ, Xia Y, Liao WY, y Ko, J. M. Technology innovation in developing the structural health monitoring system for Guangzhou new TV tower. Structural Control and Health Monitoring. 2009;**16**(1):73-98. DOI: 10.1002/stc.303

[28] Compán V, Pachón P, Cámara M, Lourenço PB, y Sáez, A. Structural safety assessment of geometrically complex masonry vaults by non-linear analysis. The chapel of the Würzburg residence (Germany). Engineering Structures. 2017;**140**:1-13. DOI: 10.1016/j.engstruct.2017.03.002

[29] Alejo LE, Mendes N, Lourenço PB, y Martínez, G. Protecting the historic buildings of Mexico: The barrel vault of san Agustin church in Morelia. Journal of

Performance of Constructed Facilities. 2021;**35**(2):04020146. DOI: gppbvp

[30] Pierdicca A, Clementi F, Isidori D, Concettoni E, Cristalli C, y Lenci, S. Numerical model upgrading of a historical masonry palace monitored with a wireless sensor network. International Journal of Masonry Research and Innovation. 2016;**1**(1):74-99. DOI: 10.1504/IJMRI.2016.074748

[31] Barsocchi P, Bartoli G, Betti M, Girardi M, Mammolito S, Pellegrini D, et al. Wireless sensor networks for continuous structural health monitoring of historic masonry towers. International Journal of Architectural Heritage. 2021;**15**(1):22-44. DOI: 10.1080/15583058.2020.1719229

[32] Fort EH, Blanco-Carmona P, Garcia-Oya JR, Chavero FM, Carvajal RG, Serrano-Chacon AR, Mascort-Albea EJ. Wireless and Low-Power system for synchronous and Real-Time Structural-Damage assessment. IEEE Sensors Journal. 2023;**23**(12):13648-13658. DOI: 10.1109/jsen.2023.3270961

[33] James GH, Carne TG, Lauffer JP, Nord AR. Modal testing using natural excitation. In: Proceedings of the 10th International Modal Analysis Conference: February 3-7, 1992. San Diego, California, USA. Society for Experimental Mechanics; 1992. pp. 1209-1216

[34] Peeters B. System identification and damage detection in civil engineering [PhD thesis, Catholic University of Leuven]; 2000

[35] Van Overschee P, De Moor B. Subspace identification for linear systems. En Springer eBooks. 1996. DOI: 10.1007/978-1-4613-0465-4

[36] Pintelon R, Schoukens J. System Identification: A Frequency Domain Approach. 2nd ed. USA: Wiley-IEEE Press; 2012

[37] Brincker R, Zhang L, Andersen P. Modal identification from ambient responses using frequency domain decomposition. In: Proceedings of IMAC-XVIII: A Conference on Structural Dynamics: February 7-10, 2000. San Antonio, Texas, USA: Society for Experimental Mechanics; 2000. pp. 625-630

[38] Brincker R, Ventura CE, Andersen P. Damping estimation by frequency domain decomposition. In: Proceedings of IMAC-XIX: A Conference on Structural Dynamics: February 5-8, 2001. Kissimmee, Florida, USA. Society for Experimental Mechanics; 2001. pp. 698-703

[39] Rytter A. Vibrational based inspection of civil engineering structures [PhD thesis, Aalborg University]. 1993. Available from: https://vbn.aau.dk/files/18588667/Vibrational_Based_Inspection_of_Civil_Engineering_Structures

[40] Bisio I, Garibotto C, Lavagetto F, Sciarrone A. A novel IoT-based edge sensing platform for structure health monitoring. In: IEEE INFOCOM 2022—IEEE Conference on Computer Communications Workshops (INFOCOM WKSHPS). Virtual Conference; 2022. DOI: 10.1109/infocomwkshps54753.2022.9797953

[41] Zou Z, Bao Y, Li H, Spencer BF, Jinping O. Embedding compressive sensing-based data loss recovery algorithm into wireless smart sensors for structural health monitoring. IEEE Sensors Journal. 2015;**15**(2):797-808. DOI: 10.1109/jsen.2014.2353032

[42] Zonzini F, Dertimanis V, Chatzi E, Marchi LD. System identification at the extreme edge for network load reduction in vibration-based monitoring. IEEE Internet of Things Journal.

2022;**9**(20):20467-20478. DOI: 10.1109/jiot.2022.3176671

[43] Mobile IoT Deployment Map. Internet of Things. 2023 https://www.gsma.com/iot/deployment-map/

[44] NB-IoT & LTE-M. Summary—GSA. 2023. Available from: https://gsacom.com/paper/nb-iot-lte-m-march-2023-summary/

[45] 1NCE Precios. 1NCE. 2023. Available from: https://1nce.com/es-es/1nce-connect/10-euros-por-10-anos

[46] Zini G, Betti M, Bartoli G, Chiostrini S. Frequency vs time domain identification of heritage structures. Procedia Structural Integrity. 2018;**11**:460-469. DOI: 10.1016/j.prostr.2018.11.115

[47] Saisi A, Terenzoni S, Ruccolo A, Gentile C. Safety of the architectural heritage: Structural assessment of the Zuccaro's Tower in Mantua. En Springer eBooks. 2019. pp. 2422-2430. DOI: 10.1007/978-3-319-99441-3_260

[48] Bayraktar A, Altunişik AC, Birinci F, Sevim B, Türker T. Finite-element analysis and vibration testing of a two-span masonry arch bridge. Journal of Performance of Constructed Facilities. 2010;**24**(1):46-52. DOI: 10.1061/(ASCE)CF.1943-5509.0000060

[49] Pepi C, Gioffrè M, Comanducci G, Cavalagli N, Bonaca A, Ubertini F. Dynamic characterization of a severely damaged historic masonry bridge. Procedia Engineering. 2017;**199**:3398-3403. DOI: 10.1016/j.proeng.2017.09.579

[50] Venanzi I, Kita A, Cavalagli N, Ierimonti L, Ubertini F. Earthquake-induced damage localization in an historic masonry tower through long-term dynamic monitoring and FE model calibration. Bulletin of Earthquake Engineering. 2020;**18**(5):2247-2274. DOI: 10.1007/s10518-019-00780-4

[51] Standoli G, Salachoris GP, Masciotta MG, Clementi F. Modal-based FE model updating via genetic algorithms: Exploiting artificial intelligence to build realistic numerical models of historical structures. Construction and Building Materials. 2021;**303**:124393. DOI: 10.1016/j.conbuildmat.2021.124393

[52] Costa C, Arêde A, Costa A, Caetano E, Cunha Á, Magalhães F. Updating numerical models of masonry arch bridges by operational modal analysis. International Journal of Architectural Heritage. 2015;**9**(7):760-774. DOI: 10.1080/15583058.2013.850557

[53] Costa C, Ribeiro D, Jorge P, Silva R, Arêde A, Calçada R. Calibration of the numerical model of a stone masonry railway bridge based on experimentally identified modal parameters. Engineering Structures. 2016;**123**:354-371. DOI: 10.1016/j.engstruct.2016.05.044

[54] Magalhães F, Cunha Á. Explaining operational modal analysis with data from an arch bridge. Mechanical Systems and Signal Processing. 2011;**25**(5):1431-1450. DOI: 10.1016/j.ymssp.2010.08.001

[55] Kortiš J, Daniel Ľ, Farbák M, Maliar L, Škarupa M. Operational modal analysis of the cable-stayed footbridge. Civil and Environmental Engineering. 2017;**13**(2):92-98. DOI: 10.1515/cee-2017-0012

[56] Diord S, Magalhães F, Cunha Á, Caetano E. High spatial resolution modal identification of a stadium suspension roof: Assessment of the estimates uncertainty and of modal contributions. Engineering Structures. 2017;**135**:117-135. DOI: 10.1016/j.engstruct.2016.12.060

[57] Mostafaei H, Ghamami M, Aghabozorgi P. Modal identification of concrete arch dam by fully automated operational modal identification. Structure. 2021;**32**:228-236. DOI: 10.1016/j.istruc.2021.03.028

[58] Avci O, Alkhamis K, Abdeljaber O, Alsharo A, Hussein M. Operational modal analysis and finite element model updating of a 230 m tall tower. Structure. 2022;**37**:154-167. DOI: 10.1016/j.istruc.2021.12.078